The Original Accident

Father, forgive them: for they know not what they do.

Christ, Luke 23:34

The Original Accident

Paul Virilio

Translated by Julie Rose

polity

Polity Press
65 Bridge Street
Cambridge CB2 1UR, UK

Polity Press
350 Main Street
Malden, MA 02148, USA

ISBN-10: 0-7456-3613-6
ISBN-13: 978-07456-3613-6
ISBN-10: 0-7456-3614-4 (pb)
ISBN-13: 978-07456-3614-4 (pb)

A catalogue record for this book is available from the British Library.

Typeset in 12 on 14 pt Bembo
by Servis Filmsetting Limited, Manchester

For further information on Polity, visit our website: www.polity.co.uk

Ouvrage publié avec le concours du Ministère français chargé de la culture – Centre National du Livre.

Published with the assistance of the French Ministry of Culture – National Centre for the Book.

This book is supported by the French Ministry for Foreign Affairs, as part of the Burgess programme headed for the French Embassy in London by the Institut Français du Royaume-Uni.

Contents

Part I

1 Caution

'One feature, the most distinctive of all, pits contemporary civilization against those that have preceded it: *speed*. The metamorphosis occurred in the space of a single generation,' the historian, Marc Bloch, noted in the 1930s.

This situation involves a second feature in turn: *the accident*. The gradual spread of catastrophic events not only affects the reality of the moment but causes anxiety and anguish for generations to come.

From incidents to accidents, from catastrophes to cataclysms, everyday life has become a kaleidoscope where we endlessly bang into or run up against what crops up, *ex abrupto*, out of the blue, so to speak . . . And so, in this broken mirror, we need to learn how to clearly make out what crops up more and more frequently and, more to the point, more and more rapidly, in an untimely fashion, perhaps even simultaneously.

Faced with this state of affairs in an accelerated temporality that affects customs and moral standards and art every bit as much as the politics of nations, one thing stands out as being of the utmost urgency: to expose *the accident in Time*.

Turning on its head the threat of the unexpected, the surprise, becomes a subject for a thesis and the natural disaster, the subject of an exhibition within the framework of instantaneous telecommunications.

As Paul Valéry explained in 1935: 'In the past, when it came to novelty, we had hardly ever seen anything but solutions to problems or answers to questions that were very old, if not age-old . . . *But novelty for us now consists in the unprecedented nature of the questions themselves, and not the solutions, in the way these questions are asked and not the answers.* Whence the general impression of powerlessness and incoherence that rules our minds.'[1]

This admission of powerlessness in the face of the surging up of unexpected and catastrophic events forces us to try to reverse the usual trend that exposes us to the accident in order to establish a new kind of museology or museography: one that would now entail exposing the accident, all accidents, from the most banal to the most tragic, from natural catastrophes to industrial and scientific disasters, without avoiding the too often neglected category of the happy accident, the stroke of luck, the *coup de foudre* or even the *coup de grâce*!

Today, thanks to television, 'what survives is reduced to the event-instant, progress of all kinds converging on an inescapable problem which is the problem of perception and image.'[2]

Apart from the historic terrorist attack of 11 September 2001 and its broadcasting on a continuous loop on the television screens of the entire world, two recent events deserve to come in for some harsh analysis on this score. On the one hand, we have the revelation, *sixteen years too late*, of the damage done to eastern France through contamination from Chernobyl, about which those running the services tasked with sounding the alert in France declared in April, 1986: *'If we do detect anything, it will just be a purely scientific problem.'* On the other hand, we have the very recent decision of the Caen Memorial Peace Museum to import from the United States, as a symbolic object, an *atomic bomb* – an H-bomb – emblematic of the 'balance of terror' during the Cold War between East and West.

Apropos, and reworking the dismissive remark of the French experts who covered up the damage done by the Chernobyl accident, we might say: *'If we exhibit an atom bomb, it will just be a purely cultural problem,'* and on that note, throwing open the doors of the first Museum of Accidents.

They say *invention is merely a way of seeing*, of reading accidents as signs and as opportunities. If so, then it is merely high time we opened the museum to what crops up impromptu, to that 'indirect production' of science and the technosciences constituted by disasters, by industrial or other catastrophes.

According to Aristotle, 'the accident reveals the substance.' If so, then invention of the 'substance' is equally invention of the 'accident'. The shipwreck is consequently the 'futurist' invention of the ship, and the air crash the invention of the supersonic airliner, just as the Chernobyl meltdown is the invention of the nuclear power station.

Let's take a look now at recent history. While the twentieth century was the century of great exploits – such as the moon landing – and great discoveries in physics and chemistry, to say nothing of computer science and genetics, it would seem, alas, only logical that the twenty-first century, in turn, reap the harvest of this hidden production constituted by the most diverse disasters, *to the very extent that their repetition has become a clearly recognizable historical phenomenon.*

On this score, let's hear it again from Paul Valéry: 'The tool is tending to vanish from consciousness. We commonly say that its function has become automatic. What we should make of this is the new equation: consciousness only survives now as awareness of accidents.'[3]

This admission of failure then leads to a clear and definitive conclusion: 'All that is capable of being resumed and repeated is fading away, falls silent. *Function only exists outside consciousness.*'

Given that the declared objective of the Industrial Revolution of the eighteenth century was precisely the repetition of standardized objects (machines, tools, vehicles, etc.), in other words, famously *incriminated* substances, it is only logical today to note that the twentieth century did in fact swamp us with *mass-produced accidents one after the other*, from the sinking of the *Titanic* in 1912 up to the Chernobyl meltdown in 1986, to say nothing of the Seveso chemical plant disaster of 1976 or of the Toulouse fertilizer factory disaster of 2001.

And so serial reproduction of the most diverse catastrophes has dogged the great discoveries and the great technical inventions like a shadow, and, unless we accept the unacceptable, meaning allow the *accident* in turn to become *automatic*, the urgent need for an 'intelligence of the crisis in intelligence' is making itself felt, at the very beginning of the twenty-first century – an intelligence which *ecology* is the clinical symptom of, anticipating the imminent emergence of a philosophy of post-industrial *eschatology*.

Let's accept Valéry's postulate: if consciousness only survives now as *awareness of accidents*, and if nothing functions except outside consciousness, the loss of consciousness about accidents as well as major disasters would not only amount to unconsciousness but to madness – the madness of deliberate blindness to the fatal consequences of our actions and our inventions. I am thinking in particular of genetic engineering and the biotechnologies. Such a situation would then mean embracing the swift reversal of *philosophy* into its opposite – in other words, the birth of *philofolly*, a love of what was repressed as radically unimaginable, unthinkable, whereby the insane nature of our acts would not only stop consciously worrying us, but would thrill us and captivate us.

After the *accident in substances*, we would see the fatal emergence of the *accident in knowledge*, which computer science could well be a sign of, due to the very nature of its indisputable 'advances' but also, by the same token, due to the nature of the incommensurable damage it does.

In fact, if 'the accident is the appearance of a quality of something that was hidden by another of its qualities,'[4] then the invention of industrial accidents in (land, sea, air) transport or of post-industrial accidents, in the fields of computers and genetics, would be the appearance of a quality too long hidden by the poor progress of 'scientific' knowledge compared to the sheer scope of 'spiritual and philosophical' knowledge, of the *wisdom* accumulated over centuries throughout the history of civilizations.

And so, the havoc wreaked by secular or religious ideologies, peddled by totalitarian regimes, is about to be outstripped by that wreaked by thought technologies that are likely to end, if we are not careful, in MADNESS, in an insane love of excess, as the suicidal nature of certain contemporary acts would tend to bear out, from Auschwitz right up to the military concept of Mutually Assured Destruction (MAD), to say nothing of the 'imbalance of terror' kickstarted in New York in 2001 by the suicide bombers of the World Trade Center.

Indeed, not to use weapons, military tools, any more but simple air transport vehicles to destroy buildings, and being prepared to die in the process, is to set up a fatal confusion between the terrorist attack and the accident and to use the 'quality' of the deliberate accident to the detriment of the quality of the aeroplane, as well as the 'quantity' of innocent lives sacrificed, thereby exceeding the bounds once set by ethics, religious or philosophical.

Actually, the *imperative of responsibility* for the generations to come requires that we now expose accidents along with the frequency of their industrial and post-industrial repetition.

This is the whole point of the exhibition at the Fondation Cartier pour l'art contemporain as well as its avowed aim. A test run or, more precisely, a prefiguration of a future Museum of the Accident, the exhibition aims first and foremost to take a stand against the collapse of ethical and aesthetic landmarks, that loss of meaning we so often witness now as victims much more than as actors.

After the exhibition on speed also organized by the Fondation Cartier over ten years ago, at Jouy-en-Josas, the exhibition *Ce qui arrive*, from the Latin *accidens* (unknown quantity in English) hopes to act as a counterpoint to the outrages of all stripes that we are swamped with on a daily basis by the major media outlets, that museum of horrors that no one seems to realize precedes and accompanies the escalation of even bigger disasters.

In fact, as one witness to the rise of nihilism in Europe puts it: 'The most atrocious act is easy when the way leading up to it has been duly cleared.'[5]

By gradual habituation to insensitivity and indifference in the face of the craziest scenes, endlessly replayed by the entertainment markets in the name of some so-called freedom of expression that has morphed into the *freeing up of expressionism*, or even into an academicism of horror, we are succumbing to the ravages of a programming of outrageousness at all costs that leads, not to meaninglessness any more, but to the selling of terror and terrorism as heroism.

Much as the official art of the nineteenth century went out of its way to glorify the great battles of the past in its salons and wound up, as we know, in the mass slaughter of Verdun, at the very dawn of the twenty-first century, we look on, gobsmacked, at the attempts to promote artistic torture, aesthetic self-mutilation and suicide as an artform.[6]

It is, in the end, in order to escape this overexposure of the public to horror that the Fondation Cartier pour l'art contemporain agreed to hold the exhibition, 'Unknown Quantity', organized by myself as an event aimed above all at keeping its distance from the outrages of every stripe with which current events are riddled.

Designed to raise the issue of the unexpected and of the lack of attention to major hazards, the exhibition manifesto endeavoured above all to pay homage to discernment, to *preventive intelligence*, at a time when threats of triggering a *preventive war* in Iraq abounded.

2 The Invention of Accidents

Creation or collapse, the accident is an unconscious oeuvre, an *invention* in the sense of uncovering what was hidden, just waiting to happen.

Unlike the 'natural' accident, the 'artificial' accident results from the innovation of a motor or of some substantial material. Whether the sinking of the *Titanic* or the eruption of the Chernobyl nuclear power station – emblematic catastrophes of the past century – the issue raised by the accidental event is not so much that of an iceberg surging up in the North Atlantic on a certain night in 1912, or that of a divergent nuclear reactor on a certain day in 1986. The issue is the building of an 'unsinkable' ocean liner or the setting up of an atomic power station close to residential zones.

In 1922, for instance, when Howard Carter stumbled across Tutankhamun's sarcophagus in the Valley of the Kings, he literally invented it. But when the Soviet 'liquidators' covered the faulty Chernobyl reactor with a different kind of 'sarcophagus', *they invented the major nuclear accident*, and this, only a few years after the one that had occurred at Three Mile Island in the United States.

So, just as Egyptology is one of the disciplines of historical discovery, in other words, of *archaeological invention*, analysis of the industrial accident ought to be seen as a 'logical art' or, more precisely, as an archaeotechnological invention.

An *art brut* in every sense of the term, but one we can't look at solely as an exception or, from the preventive angle, as a 'precautionary principle' alone. It has to be seen equally as a major work of unconscious scientific genius, the fruit of Progress and of the labour of mankind.

Note, on top of this, that if techniques are always streets ahead of the mentality of *users*, who need several years to get used to a new technology, they are also streets ahead of the mentality of *producers*, those engineers busily engineering the invention of engines – so much so that the mechanical unconscious once flagged by psychoanalysis here proves its validity as a *proof through absurdity* of the fatal recklessness of scientists when it comes to knowing about major risks.

'There is no science of the accident,' Aristotle cautioned a long time ago. Despite the existence of risk studies which assess risks, there is no *accidentology*, but only a process of fortuitous discovery, archaeotechnological invention. To invent the sailing ship or steamer is *to invent the shipwreck*. To invent the train is *to invent the rail accident* of derailment. To invent the family automobile is to produce the *pile-up* on the highway.

To get what is heavier than air to take off in the form of an aeroplane or dirigible is *to invent the crash*, the air disaster. As for the space shuttle, *Challenger*, its blowing up in flight in the same year that the tragedy of Chernobyl occurred is the *original accident* of a new motor, the equivalent of the first shipwreck of the very first ship.

An indirect invention of the breakdown of computer (or other) systems, look at the economic upheaval in the financial markets when suddenly, with the stockmarket crash, the hidden face of the economic sciences and technologies of automated dealing in values rears up, like the iceberg before the *Titanic*, only on Wall Street, in Tokyo and in London.

And so, if, for Aristotle some little time ago and for us today, *the accident reveals the substance*, this is in fact because WHAT CROPS UP (*accidens*) is a sort of analysis, a techno-analysis of WHAT IS BENEATH (*substare*) any knowledge.

It follows that fighting against the damage done by Progress above all means uncovering the hidden truth of our successes in this accidental revelation – in no way apocalyptic – of the incriminated substances.

Whence the urgent need, at the threshold of the third millennium, for public recognition of this type of innovation that comes and feeds off every technology, as the twentieth century never ceased stunningly demonstrating.

On this score, too, *political ecology* cannot long go on sweeping under the carpet the *eschatological* dimension of the calamities caused by the positivist ideology of Progress.

So the *dromologue*, or, if you like, the analyst of the phenomena of acceleration, is consistent in thinking that if speed is responsible for the exponential development of the *artificial accidents* of the twentieth century, it is also every bit as responsible for the increased impact of *ecological accidents* (the sundry instances of pollution of the environment) as, let's say, the *eschatological calamities* that are looming with the very recent discoveries of genomics and biotechnologies.[1]

Once upon a time the local accident was still precisely situated – as in the North Atlantic for the *Titanic*. But the global accident no longer is and its fallout now extends to whole continents, anticipating the *integral accident* that is in danger of becoming, tomorrow or the day after, our sole habitat, the havoc wreaked by Progress then extending not only to the whole of geophysical space, but especially to timespans of several centuries, to say nothing of the dimension *sui generis* of a 'cellular Hiroshima'.

Actually, if the substance is *absolute and essential* (to science) and if the accident is *relative and contingent*, we can now identify the 'substance' at the *beginning* of specific fields of knowledge and the 'accident' at the *end* of the philosophical intuition that Aristotle and a few others pioneered.

Far from urging some 'millenarian catastrophism', there is no question here of making *a tragedy* out of an accident with the aim of scaring the hordes as the mass media so often do, but only of finally taking accidents *seriously.*

Along the lines of the work of someone like Freud on our relationship to death and the impulse towards it, it is now a matter of scrupulously examining *our relationship to the end*, to all ends, in other words to finiteness.

'Accumulation puts an end to the impression of chance,' wrote Sigmund Freud, some time between 1914 and 1915. Indeed, after the twentieth century and the sudden *capitalization of tragedies and catastrophes of all kinds*, we really should draw up the bankruptcy report on a technoscientific Progress that the nineteenth-century positivists were so proud of.

Since those days the serial production of the wizards of business has literally *industrialized the artificial accident*, an accident whose once *artisanal* character most often expressed itself discreetly, even while *natural* accidents took on a cataclysmic dimension all of their own, with the exception of wars of annihilation.

If we take the realm of private car ownership, for example, the way the carnage on the highways has become commonplace is Freudian proof that the accumulation of traffic accidents largely puts an end to 'chance' – and the multiple security systems our vehicles are equipped with don't alter this fact one iota: in the course of the twentieth century, the accident became a heavy industry.

But let's get back to this technoanalysis revelatory of 'substance' – in other words, what lies beneath technicians' knowledge. Techniques are always streets ahead of the mentalities of competent personnel in the area of innovation, as the essayist, John Berger, likes to claim, in any case ('*In every creation*, whether it involves an original idea, a painting or a poem, *error always sits alongside skill*. Skill is never presented on its own; *there is no skill, no creative talent, without error*').[2] But this is because the accident is inseparable from *the speed with which it unexpectedly surges up*. And so this 'virtual speed' of the catastrophic surprise really should be studied instead of merely the 'actual speed' of objects and engines fresh off the drawing board.

Just as we need to protect ourselves (at any cost) from *excess in real speed* by means of breaks and automated safety systems, we have to try and protect ourselves from *excess in virtual speed*, from what unexpectedly happens to 'substance', meaning to what lies *beneath* the engineer's awareness as producer.

This is the 'archaeotechnological' invention itself, the discovery evoked above.

In his *Physics*, Aristotle remarks at the outset that it is not Time as such that corrupts and destroys, but what crops up (*accidens*). So it is indeed *the passage of Time*, in other words the speed with which they crop up that achieves the ruin of all things, every 'substance' being, in the end, *a victim of the accident in the traffic circulation of time.*

That being the case, it's all too easy to imagine the havoc wreaked by the accident in Time, with the instantaneity of the *temporal compression* of data in the course of globalization, and the unimaginable dangers of the synchronization of knowledge.

And so, the 'imperative of responsibility' evoked by Hans Jonas really ought to be applied, in the first place, to the need for a new intelligence or understanding of the *production of accidents*, this reckless industry that the 'materialist' scientist refuses to think about, even though the 'military-industrial complex' bombarded us, throughout the entire past century, with the sudden militarization of the sciences, most notably, the fatal invention of weapons of mass destruction and a thermonuclear bomb capable of extinguishing all life on the planet.[3]

In fact, the *visible speed* of the substance – that of the means of transport, of computing, of information – is only ever the tip of the iceberg of the *invisible speed* of the accident. This holds true just as much for road traffic as for the traffic of values.

If you need convincing, all you have to do is look at the very latest stock exchange crashes, the successive burstings of the speculative bubbles of the single market in a financial system that is now interconnected or has gone on-line.

Faced with this state of affairs, very largely catastrophic for the very future of humanity, we have no choice but to take stock of the urgent need for making perceptible, if not visible, the speed with which accidents surge up, plunging history into mourning.

To do this, apart from searching in vain for some black box capable of revealing the parameters of the contemporary disaster, we have to try *as fast as possible* to define the flagrant nature of disasters peculiar to new technologies. And we have to do this using scientific expertise, of course, but also a philosophical and cultural approach that would wash its hands of the *promotional expressionism* of the promoters of materials, since, as Malraux said, 'culture is what made man something more than an accident in the Universe.'

3 The Accident Argument

> Progress and disaster are two sides of the same coin.
>
> Hannah Arendt

Lately, as though an accident was now an *option*, a privilege granted to chance to the detriment of error or the desire to do harm, the accident argument has become one of the mass media's pet themes, flagging, by this very fact, the confusion now creeping in between sabotage and breakdown, on the one hand, and between the suicide bombing mission and the industrial or other accident, on the other.

Actually, the unprecedented increase in the number of catastrophes since the start of the twentieth century and right up to the present day when, for the first time, 'artificial' accidents have outstripped 'natural' accidents, makes everyone aware that they have to choose, meaning opt, for one or the other version of whatever calamity might be under way. Whence the weirdly academic expression: the accident argument.

And so, since the end of the past century, disruption – fracture – has gradually become a matter of conjecture and no longer, exactly, an unexpected surprise, causing the very term 'accident' to shed its classical philosophical meaning, which it has enjoyed since Aristotle.

Suddenly, an accident is no longer unexpected, it turns into a rumour, a priori scandalous, in which the presupposition of a fault tends to outpace anything involuntary or, conversely, the near certainty of the will to do harm is covered up in the overriding concern not to provoke panic.

We might note, here, the presumption of guilt immediately heaped on anyone refusing to buy the *official argument* for a fault or an accident and who favours instead a version completely at odds with the one touted by the powers that be.

In any case, as soon as the catastrophic event emerges in its 'terrorist' dimension, the term 'argument' is swiftly dropped for the (police) term invoking the lead or line of enquiry following a criminal act.

This semantic blurring illustrates pretty clearly the building confusion between the 'genuine' accident occurring unexpectedly to a substance and the indirect strategy of a malicious act completely typical but disdaining anything as obvious as openly declared hostility – something the rules of classic warfare still required not so long ago *(frighten, certainly, but avoid at all costs releasing a terror that is unspeakable and counterproductive for its anonymous authors)*, in a society where the screen has become a substitute for the battlefield of the great wars of the past.

The general trend towards negation of any terrorist attack – a new type of negationism that is emerging – is part and parcel, now, of the importance of the *corporate image* of any country or nation open to the cross-border tourism industry that is constantly growing thanks to the low cost of transcontinental transport.

Whence the gravity of the New York attack, which calls into question not only the United States's status as a sanctuary, but also the boom in the major airlines and the liberalization of tourist flows, to say nothing of the catastrophic impact of the collapse of the Twin Towers on the comprehensive insurance market.[1]

From now on, faced with the ubiquity of risk, often even of a major risk of disaster for humanity, the issue of *fear management* becomes crucial once more.

To paraphrase a like-minded writer, we might even assert today that: 'If knowledge can be shown as a sphere whose volume is endlessly expanding, *the area of contact with the unknown is growing out of all proportion.*'[2]

By replacing the geometric term *sphere* with the spatio-temporal term *dromosphere*, we can't help but come to the conclusion that, if the speed at which the unknown has been growing expands or intensifies fear, this alarm in the face of the final end of humanity of which the ecology movement represents an early warning sign, then that fear is set to increase even further in the twenty-first century, in anticipation of one last movement emerging, an *eschatology* movement, this time, that would be concerned with stockpiling the dividends of terror.

The abrupt undermining of the *substantial war* that derived from politics via hyperterrorism, this *accidental war* that no longer speaks its name, also undermines politics – and not only traditional party politics.

Whence the alternation not so much between the traditional left and right any more than between politics and the media, in other words, this information managing (generating) capability that is gearing up to invade the imaginary of populations held in thrall by a proliferation of screens that perfectly typifies the globalization of 'affects' – this sudden synchronization of collective emotions greatly favouring *the administration of fear.*

To administer fear in order to manage security and civil peace or, conversely, to administer fear to win a civil war – that is indeed the alternative that today characterizes the psychopolitics of nations.

As you can easily imagine, anxiety and doubt about the origin of an accident are part and parcel of this underhand administration of emotions; so much so that, in the near future, the Ministry of War could well be shunted aside for the 'Ministry of Fear' run by the movie industry and the mass media as integral parts of the *audiovisual continuum* now replacing the public space of our daily lives.

This explains the strategic urgency of maintaining uncertainty about the origin of each and every 'accident' for as long as possible since the declared enemy and official hostilities between the old states and governments have been put paid to now by the *anonymous attack* and the sabotage of daily routine, in public transport or in business firms as at home.

By way of a convincing example of this transmutation in 'politics-as-spectacle', we might cite the Hollywood blockbuster of 2002, an adaptation of Tom Clancy's 1991 novel about terrorism, *The Sum of All Fears*, sponsored by the US Department of Defense with the direct involvement of the CIA and its veteran agent, Chase Brandon, who is not afraid to claim, for his personal use, a phrase from the Gospel of St John: 'And ye shall know the truth, and the truth shall make you free' (John 8:32).[3]

In the winter of 2001, the US Defense Department announced the quiet, not to say furtive, creation of a new Office of Strategic Influence (OSI). Placed under the control of the Under-Secretary of Defense charged with politics, Douglas Feith, this information operation, a veritable 'Disinformation Department', was tasked with the diffusion of false information designed to influence 'the hearts and minds' of a terrorist enemy, itself just as diffuse . . . a *strategy of deception* from which the media of countries allied to the United States would obviously not be exempt.

Very swiftly, though, as you might expect, the Secretary of Defense, Donald Rumsfeld, was to denounce a project designed to manipulate public opinion in enemy or allied states indiscriminately. At the end of February 2002, the OSI affair was officially canned.

Now there's a fine example for you of an information accident, in other words, of brainwashing designed to sow doubt about the truth of the facts, thereby creating anxiety over diffuse threats whereby any disturbance in perception of events always reinforces the anguish of the masses.

Suicide bombing or *accident? Information* or *disinformation?* From now on, no one really knows.

In this example, which is just one among many, privileging the accident remains (as long as is necessary) the preferred option of the administration of this public fear that has nothing to do with the private fear of individuals, since the intended aim is above all *emotional* control to psychopolitical ends.

Confronted by this chain of media events, each one more catastrophic than the last (the anthrax virus, the threat of a radiological bomb, and so on), it is surely appropriate to ask ourselves about the dramatization that has been taking place since the beginning of the twenty-first century, in New York, Jerusalem and Toulouse as well as Karachi and elsewhere.

The first objective of this new dramatic art is: to never break the chain of emotion set in train by catastrophic scenes.

Whence this crescendo close to the end of a media show kicked off by Greek tragedy at the same time as Athenian democracy. In fact, for the historian of Antiquity, as for the modern philosopher, the tragic chorus is the city itself, where the future is played out between the menace of a single person and the war of each against all; this stasis that democracy must protect itself from every bit as much as from the lone tyrant.

With the globalization of the real time of telecommunications, as the new century gets under way, the public stage of the theatre of our origins gives way (and how!) to the public screen, on which the 'people's acts' are played out, this liturgy where repeat catastrophes and cataclysms have the role of some *deus ex machina*, if not of the oracle announcing the horrors to come and denouncing, thereby, the abomination of the destiny of peoples.

With television, which allows hundreds of millions of people to see the same event at the same moment in time, we are finally living through the same kind of dramatic performance as at the theatre in days not long gone. From that point on, as Arthur Miller explains, 'there is no difference between politics and show business anymore; it is the performance that persuades us that the candidate is sincere.'[4]

This has reached the stage where the people's elected representative is scarcely more than a living audimat measuring audience ratings! To maintain the illusion, to keep up the play being performed before your disbelieving eyes, at all costs, that is the objective – the tele-objective – of the contemporary mass media in the age of synchronization of opinions. Anything that destroys this collective 'harmony' must be mercilessly censured.

Since 11 September 2001, as we've all been able to observe, media coverage of acts of violence has everywhere expanded. From local delinquency to the global hyperviolence of terrorism, no one has managed to escape this escalating extremism for long. And the accumulation of felonies of a different nature has little by little given the impression that all forms of protection collapsed at the same time as the World Trade Center.[5]

And so this dramatic portrayal has created, in televiewers, a twin fear, a stereo-anxiety. Alarm over public insecurity has been topped up with fear of the images of 'audiovisual' insecurity, bringing about a sudden highlighting of domestic terror, designed to intensify collective anguish. 'We are living off the echo of things and, in this upside down world, it is the echo that gives rise to the cry,' Karl Kraus once observed.[6]

This mute cry of the hordes of the absent, all present at the same moment in front of their screens contemplating disaster, stunned, is not without repercussions. The results of the French elections of 21 April 2002 prove the point abundantly, for 'it is not so much the event now as the anaesthesia that makes it possible and bearable that offers us explanations.'[7]

The sudden stereoscopic highlighting of the event, accident or attack, thus well and truly amounts to the birth of a new type of tragedy, one not only audiovisual, but binocular and stereophonic, in which the perspective of the real time of synchronized emotions produces the submission of consciences to this 'terrorism in evidence' – that we see with our own eyes – that further enhances the authority of the media.

ACCIDENT or ATTACK? From now on, uncertainty rules, the mask of the Medusa is forced on everyone thanks to Minerva's helmet or, rather, this visual headset that endlessly shows us the repetition (in a mirror) of a terror we are utterly fascinated by.

On 6 May 1937, as the afternoon drew to a close, the dirigible *Hindenburg* caught fire above Lakehurst not far from New York. It was the first great aeronautical catastrophe of the twentieth century and it counted thirty-four victims.

A young journalist commented on the event, live, on radio. His name was Orson Welles, almost the same name as that of the novelist who, some thirty years earlier, in 1908, described the bombing of New York by German dirigibles in his book, *The War in the Air*.[8]

Within thirty interminable seconds, the ocean liner of the air was blazing away like a torch in front of the news cameras and the thousands of onlookers waiting for the zeppelin to land.

Accident or sabotage? Three commissions of inquiry tried to determine the causes of this spectacular tragedy, in those days of political woes. The final verdict very quickly favoured the accident argument, by the same token bringing about the final abandonment of passenger transportation by this type of air carrier.

There, too, without radiophony and the newsreel cinema of Fox-Movietone, this major accident would not have had the mythical impact it has had – not being on anything like the scale, for instance, of the 1,500 victims of the *Titanic* twenty-five years earlier.

Similarly, this event, dire as it was for future relations between the United States and Nazi Germany, would not have found its place in history without the association of the genius of Orson Welles and that of Herbert George Wells – at the exact moment when, if not *The War of the Worlds*, also made into a movie by Orson Welles, the Second World War was about to break out and set the skies ablaze over Hiroshima and Nagasaki as its grande finale.

Now that they're ginnying up not only to relaunch dirigible transportation, but to fly transatlantic planes that can seat 500 or even 1000, the question that must now be asked is where the qualitative (if not quantative) progress lies in such loopy overkill.

Aviation accident or sabotage? The question must be asked, over and over again, unless we decide that, in the end, the fact of wanting to fly thousands of passengers at the same time in one and the same air carrier is already an accident or, more exactly, sabotage of prospective intelligence.

4 The Accident Museum

A society that unthinkingly privileges the present, real time, to the detriment of past and future, also privileges accidents. Since, at every moment, everything happens, most often unexpectedly, a civilization that implements immediacy, ubiquity and instantaneity, stages accidents and disasters.

Confirmation of this fact is provided for us, moreover, by the insurance companies, in particular by the Sigma study recently conducted on behalf of Swiss Re, the second biggest re-insurer in the world.

Recently made public, this study, which has listed, every year for the last twenty years, technical disasters (explosions, fires, acts of terrorism, and so on) and natural disasters (floods, earthquakes, hurricanes, and so on), only takes into account the set of disasters exceeding 35 million US dollars in damage.

'For the first time,' the Swiss analysts note, 'since the 1990s, a period when the damage due to natural catastrophes was greater than technical damage, the trend is the reverse, with technical damage at 70 per cent.'[1]

Proof, if proof were needed, that far from promoting quietude, our industrialized societies have, over the course of the twentieth century, intensified anxiety and increased major risks, and this is not to mention the recent proliferation of weapons of mass destruction. Whence the urgent necessity of reversing a trend that consists in exposing us to the most catastrophic accidents deriving from technoscientific genius, in order to kick-start the opposite approach which would consist in exposing the accident – exhibiting it – as the major enigma of modern progress.

Even though certain car manufacturers conduct more than 400 crash tests a year in a bid to improve the safety of their vehicles, the television networks never cease inflicting the road death toll on us – to say nothing of the endless reruns of the tragedies that make current affairs so dismal. And so it is merely high time that ecological approaches to the various forms of pollution of the biosphere are finally supplemented by an eschatological approach to technical progress, to this finiteness without which dear old globalization itself risks becoming a life-size catastrophe.

A catastrophe at once natural and artificial, a general catastrophe, one no longer specific to this or that technology or to this or that part of the world, one that would far outpace the disasters currently covered by the insurance companies of which the long-term tragedy of Chernobyl remains the patent symbol.

In order to avoid shortly inhabiting the planetary dimensions of an integral accident, one capable of integrating a whole heap of incidents and disasters through chain reactions, we must start right now building, inhabiting and thinking through the laboratory of cataclysms, the museum of the accident of technical progress. This is the only way to avert the sudden springing up, in the near future, following the accident in substances – revealed by Aristotle – of the accident in all knowledge, a full-scale philosophical accident which genetic engineering, in the wake of atomic engineering, now portends.

Whether we like it or not, globalization is today the fatal trademark of finiteness. Paraphrasing Valéry, we could say with some confidence that, 'the time of the finite world is beginning.' It is urgent we grasp that knowledge marks the finiteness of man, exactly as ecology marks the finiteness of man's geophysical environment.

At the very moment that some people are calling, by open letter to the president of the Republic of France, for the creation of a 'museum of the twentieth century' in Paris,[2] it would seem appropriate to ask ourselves not only about the historical chain of events of that fateful century, but also about the fundamentally catastrophic nature of those events.

If 'time is the accident of accidents,'[3] then the history museums already in fact anticipate the museum of this integral accident which the twentieth century paved the way for on the pretext of some or other scientific revolution or ideological liberation.

Every museology requiring a museography, the question of how to show the havoc wreaked by progress is left hanging but must be asked as a crucial aspect of the project.

Here, we have to say that, far from the history books and chronicles of the press, radio, followed by the cinema newsreels and especially television, foreshadowed the historic laboratory we are talking about.

Indeed, since cinema is time exposing itself as the sequences scroll past, with television, it is clearly the pace of its 'cross-border' ubiquity that shatters the history that is in the making before our very eyes.

And so, general history has been hit by a new type of accident, the accident in its perception as visibly present – a 'cinematic' and shortly 'digital' perception that changes its direction, its customary rhythm, the rhythm of the ephemerides or calendars – in other words the pace of the long time-span, promoting instead the ultra short time-span of this televisual instantaneity that is revolutionizing our vision of the world.

'With speed, man has invented new kinds of accidents. [. . .] The fate of the motorcar driver has become a matter of sheer luck,' Gaston Rageot wrote, in the 1930s.[4]

What can you say, today, of the major accident of audiovisual speed and so of the fate of the numberless hordes of televiewers? If not that thanks to it, history becomes 'accidental' through the abrupt telescoping of facts and the collision of events once successive that have become simultaneous, despite the distances and time lapses once necessary to their interpretation.

Just imagine, for instance, the probable damage done by the practice of live digital morphing to the authenticity of the testimony of history's actors.

'For a long while the movies took their cue from the other arts, now it is the visual arts that take their cue from the movies,' Dominique Païni recently lamented apropos the dominant influence of film on the conception of contemporary art.

But, in fact, history as a whole takes its cue from filmic acceleration, from this cinematic and televisual crush! This is behind the ravages in the circulation of images, the constant telescoping, the pile-up of dramatic scenes from daily life on the nightly news broadcast. If the print media have always been interested in trains that get derailed rather than those that arrive on time, with the audiovisual we are able to look on, flabbergasted, at the overexposure of accidents, catastrophes of all kinds, to say nothing of wars.

With the television image, we have looked on, live, since the end of last century, at endless overkill in the broadcasting of horror and, especially since the boom in live coverage, in the instantaneous broadcasting of cataclysms and terrorist outrages that have largely had the jump on disaster films.

Even more to the point, following the standardization of opinion that came in with the nineteenth century, we are now witnessing the sudden synchronization of emotions.

The ratings war of the television networks has turned the catastrophic accident into a scoop, not to say a fantastic spectacle sought after by all.

When Guy Debord spoke of the 'showbiz society', he forgot to mention that this adaptation of life to the screen is based on sexuality and violence; a sexuality that the decade of the 1960s claimed to liberate, whereas the real agenda involved obliterating societal inhibitions one by one, considered as they were by the situationists as so many unacceptable constraints.

One of the organizers of the Avoriaz science fiction film festival puts it perfectly: 'Death has finally replaced sex and the serial killer, the Latin lover!'

'Museum of horrors' or 'tunnel of death' television has thus gradually been transformed into a sort of altar of human sacrifices; using and abusing the terrorist stage and repeat massacres, television now plays on repulsion more than on seduction. From the so-called live death of a little Colombian girl who sank into the mud twenty years ago, right up to the execution, in September 2000, of little Mohamed Al-Dura, hit by a bullet as he lay in his father's arms, any excuse will do for feeding the fear habit once it is created.

Conversely, as you will recall, the mass media of the old Soviet Union never gave out information about accidents or attacks. Except for natural disasters that were pretty hard to sweep under the carpet, the media outlets systematically censured any breach of norms so that only the horizon of a radiant future would filter through . . . This, right up to Chernobyl.

But, speaking of censorship, liberalism and totalitarianism each had their own peculiar method for smothering the truth of the facts. For liberalism, the process already involved overexposing the televiewer to incessant replays of calamities, while totalitarianism opted for underexposure and radical concealment of anything and everything out of the ordinary. Two separate panic movements, but producing an identical result: censorship by floodlighting, fatal bedazzlement for the democratic West; and censorship by banning every divergent representation, the 'night and fog' of deliberate blindness for the dogmatic East.

Just as there exists a Richter scale for telluric catastrophes, there also exists a sort of secret scale of media-relayed catastrophes whose most obvious effect is to inspire resentment against those running the show, on the one hand, and, on the other, an effect of exemplarity that ends, when it comes to terrorism, in reproduction of the disaster, thanks to its dramatic amplification. This has reached the point where it would now be appropriate to supplement the birth of tragedy once studied by Nietzsche with an analysis of this media tragedy where perfect synchronization of the collective emotion of televiewers would play the same role as the Greek chorus of antiquity, not on the scale of the theatre of Epidaurus now, but on that life-size scale of whole continents.

This is obviously where the Accident Museum comes in. Actually the media scale of the catastrophes and cataclysms that cripple the world with grief is now so vast that it must necessarily make the magnitude of the field of perception the first phase of a new intelligence, not only that of the ecology of hazards due to pollution of the environment now, but also that of an ethology of threats in terms of brainwashing public opinion, of polluting public emotion.

This form of pollution always paves the way for intolerance swiftly followed by revenge, in other words for forms of barbarity and chaos that soon overrun human societies. This is amply demonstrated by the massacres and genocides of our day that blowback from the deadly propaganda of the 'media of hate'.

After all the waiting for the integral accident, we are now witnessing the forceps birth of a 'catastrophism' that has nothing in common (not by any means) with the pessimism of the 'millenarian' obscurantism of days gone by. But it does mean we need to be just as careful and use that Pascalian *esprit de finesse*, a sharp subtlety that the mass information outlets are so cruelly lacking in.

Indeed, since one catastrophe can hide another, if the major accident is indeed the consequence of the speed of acceleration of phenomena engendered by progress, it is merely high time, now we've negotiated the turn of the twenty-first century, that we analyse wisely what crops up, what surges out of the blue before our very eyes, leading to the overwhelming necessity, now, for *exposing* the accident.

One last example to wrap up: fairly recently, astronomers have been cataloguing and monitoring asteroids and meteorites heading for Earth.

Baptized 'geo-cruisers', these earthbound fireballs some tens of metres in diameter obviously represent a threat of collision with our planet.

The last direct hit from such a rogue object occurred over Tunguska in Siberia in 1908 and its explosion at an altitude of 8,000 metres flattened an area of over 2,000 square kilometres of forest.

To try and avert recurrence of such a cosmic catastrophe, particularly in overpopulated areas, a working group was subsequently set up. Thanks to the support of the International Astronomical Union, this team was able to come up with a scale of NEO (Near-Earth Objects) hazards known as the Torino Scale, after the Italian town in which the protocol was adopted in 1999. Running from zero to ten, this scale of cataclysms takes account of the mass, speed and predicted path of the celestial body concerned.

Five zones have now been indexed by the scientists: a white zone where there is no chance (sic) of reaching Earth; a green zone where there is only a minuscule probability of contact; a yellow zone where there already exists a probability of impact; an orange zone where this probability is significant and, lastly, a red zone where catastrophe is inevitable.[5]

An illustration in no way alarmist of the cosmic activities which the surface of the night star bears the traces of, not to mention the Barringer Crater in Arizona over a kilometre in diameter which is very popular with American tourists, this very first attempt to expose the accident to come demonstrates the urgent need of establishing, in the twenty-first century, along the lines of the famous 'cabinets de curiosités' of the Renaissance,[6] a MUSEUM OF THE ACCIDENT OF THE FUTURE.

5 The Future of the Accident

> The world of the future will be an ever more demanding struggle against the limitations of our intelligence.
>
> Norbert Wiener

Nothing lost, nothing gained. If inventing the substance means indirectly inventing the accident, the more powerful and high-performance the invention, the more dramatic the accident.

And so the awful day comes when Progress in knowledge becomes unbearable, not only due to the various forms of pollution it creates, but due to its feats, the very power of its negativity.

This was confirmed for us throughout the twentieth century, with the race for nuclear and thermonuclear weapons that are ultimately unusable and of great concern to the protagonists of deterrence, *all-out* deterrence.

The very power of atomic weaponry in fact also flags the ultimate limitation of a power that suddenly morphs into powerlessness. In this instance, the accident is the panic-stoking uselessness of this type of weaponry.

Rather than actually fighting, military commanders then engage in the imaginary of a 'wargame' that doesn't add up, where virtuality is merely the mark of the political inconsequence of nations, since the consequences no longer really matter. For they are at once too enormous to be seriously taken in and too appalling to be viably put to the test . . .

except by some lunatic, advocating the suicide attack against humanity.

On that score, let's hear what Friedrich Nietzsche had to say in his book, *The Birth of Tragedy*, written just after the Franco-Prussian War of 1870: 'A culture built on science must necessarily perish when it starts to become illogical, that is, to recoil before its [own] consequences. Our Art reflects this general crisis.'[1]

In fact, if 'in tragedy the state of civilization is suspended,'[2] the whole panoply of beneficial knowledge finds itself wiped out with it. In total war, the sudden militarization of science necessary to the presumed victory of the adversaries turns all logic and all political wisdom on its head, to the point where age-old philo-*sophy* is shunted aside by the absurdity of a philo-*folly* capable of destroying the knowledge accumulated over the course of centuries. 'Inordinately enhanced, human power then transforms itself into a cause of ruin,'[3] toppling the whole of the culture of nations into the vacuousness of causes that are lost, irremediably lost, in victory as in defeat, since we will not be able to disinvent a terrorist and sacrilegious knowledge produced by scientific intelligence.

And so, just as there are stormy patches in nature, there are stormy patches in culture and we would need a veritable 'meteorology' of invention to avert the storms of the artifice of Progress in knowledge, that genie that stokes the escalating extremism of the power of our tools and our substances and, with them, industrial and post-industrial accidents. We can't help but think here primarily of genetics and computer science, after the fallout from atomic progress, of which Chernobyl, in the wake of Hiroshima, has revealed to us the atrocious truth.

'It's amazing what those that can do anything can't do,' declared Madame Swetchine sometime in the nineteenth century.[4] This aphorism sums up perfectly the paradox of the twentieth century and its repeat revolutions, that were like so many weapons turned on the intelligibility of the world.

Today, at the very dawn of the twenty-first century, when much-vaunted globalization is nothing if not the forbidden fruit of the tree of knowledge – in other words, of the so-called 'information revolution' – the exterminator takes over from the predator, just as terrorism takes over from the original capitalism.

Since extermination is the illogical outcome of accumulation, the suicidal state is no longer exclusively psychological, associated with the mentality of a few disturbed individuals, but sociological and political. This has reached the point where the widespread accident, announced by Nietzsche, now incorporates this dimension of panic, whereby the philosophy of the Enlightenment bows down before the philosophy of magnitude. This is, in fact, the accident in *knowledge* that now rounds off the accident in *substances* deriving from technoscientific research.

In fact, if matter has three dimensions, mass, energy and information, then, after the long series of accidents in materials and energy over the past century, the time of the logical – and even biological – accident is upon us, with the teratological research of genetic engineering.

'The machine has declared war on God,' wrote Karl Kraus,[5] you might remember, as the butchery of the First World War began. But what's the state of play today, with this globalization touted by the promoters of Progress?

A fruit of the telecommunications revolution, the globalization of knowledge has not only reduced the field of human activity to nothing thanks to the synchronization of interactivity. It also triggers a historic mutation in the very notion of accident.

The local accident, precisely located here or there, has been abruptly replaced by the possibility of a global accident that will involve not only 'substances' – the substance of the world in the age of the real time of exchanges – but also the knowledge we have of reality, that vision of the world that our various branches of knowledge were, once, founded on.

And so, after the accident in substance, we are ushering in with the coming century an accident without parallel, an accident in the real, in space, in time as in substantial matter, which the cynics had no idea about but which the physicists of relativity introduced bit by bit, in the course of total war.

'Time is just an illusion,' declared Albert Einstein, during the period that divided the First World War from the Second. An accident in historical knowledge, in other words, in the perception of things, a veritable loss of the sense of reality – the fruit of a reality now spiralling off in accelerated flight, just like the galaxies in the expanding universe. Werner Heisenberg already foresaw the devastation such a loss would cause, fifty years ago, when he wrote: 'No one knows what will be real for people at the end of the wars now beginning.'[6]

In the end, after the implosion of the Cold War between East and West, globalization is above all a sort of 'voyage to the centre of the Earth' in the gloomy obscurity of a temporal compression that definitively locks down the habitat of the human race. Certain utopians were already calling this the sixth continent, though all it is is the hypercentre of our environment.

This hypercentre is at once origin and end of a world now foreclosed, where each and every one of us is endlessly pulled towards this central region, without expanse and without temporal extension. And yet it is merely the outcome, the terminal, of this acceleration of reality that crushes everyone together, all five continents and all seven seas, and especially, the nations and peoples of the planet in its entirety.

Here we have a telluric compression of the history of humanity that, despite the ecologists, no seismograph is registering the magnitude of this cataclysm wherein everything is telescoped, rammed into everything else at every instant, where all distances are reduced to nothing, obliterated by the accident of the real time of interactivity. And this earthquake affects the whole Earth, with events now nothing more than untimely and simultaneous accidents, at the surface of a celestial object crazily compressed where gravity and atmospheric pressure are further reinforced by the instantaneous synchronization of exchanges.

At this level of anxiety, ecology is less bound up with nature than with culture and the ethological catastrophes culture has in the works. In effect, with the *mise en abyme* of time ratios, lags and scales, the instantaneous elimination of any interval in the promotion of immediacy, this pollution of the distances of the life-size scale of the globe teaches us infinitely more than the pollution of the substances of nature about the calamity, the tragedy of future branches of knowledge. In the frightening compression of the far-flung extremities of a once-gigantic world towards the centre, the hypercentre of the only habitable planet in the solar system, 'Nature can have confidence in Progress; Progress will know how to avenge the affronts it has made it suffer.'[7]

By way of conclusion, let's ask ourselves three questions: Should science reassure? Should science, on the contrary, frighten? And, lastly, is science inhuman?[8]

So many inquiries that largely throw light on the famous 'crisis in Progress'; and also, every bit as much, on the crisis, in no way subsidiary, created by the recent mediatization of discoveries, this 'scientific expressionism' certain mad scientists have been frantically peddling. One such is the Italian gynaecologist, Severino Antinori, the 'Doctor Strangelove' of assisted procreation; another is university professor and cancer specialist, Friedhelm Hermann, accused, in the autumn of 1999, by a German commission tasked with tracking down laboratory fraud, of having doctored his team's results, thereby triggering a veritable 'Chernobyl of science', according to the specialist press![9]

We might recall, at this juncture, that the freedom of expression touted by the sensationalist press can never be the same as that of scientific research without sooner or later leading to the 'philofolly' of a science not only deprived of a conscience but deprived of meaning.

Atomic bomb yesterday, information bomb today and, tomorrow, genetic bomb?

In August 2001, before the National Academy of Sciences in Washington, Professor Antinori presented his project of bringing about the birth of some two hundred babies through reproductive cloning, promising the infertile 'parents' perfect children, even if it meant discarding the imperfect. What is at work here if not some kind of demiurgic insanity? Proof, if proof were needed, that when it comes to scientific issues, as elsewhere, the worst case scenario is sometimes a pretty safe bet.[10]

Radioactive fallout from Chernobyl, genetically modified organisms, reproductive human cloning following on from animal cloning – the list goes on. Scientific experts now find themselves smack-bang in the middle of controversies that are shaking up the dawning third millennium. This is behind the recent creation of agencies specializing in risk management in a bid to try and forecast the improbable or unthinkable in scientific and technical knowledge. For it is true that for some decades now, we have been confronted defencelessly by major risks that affect the biological and social balance of humanity.[11]

Looked at from this particular angle, the 'accident in knowledge' is impressive not so much in terms of the number of victims but in the very nature of the risk run.

Unlike road, rail or air accidents, that risk is no longer quantifiable and statistically predictable; it has become unqualifiable and fundamentally unpredictable. This has reached the point of entailing the emergence of an unparalleled risk, whose scope is no longer exclusively ecological, connected to the conditions of the surrounding human habitat, but eschatological, since what it attacks is the mind's ability to anticipate; in other words, it attacks rationality itself.[12]

'Ruin of the soul', wrote Rabelais about a science without a conscience. . . . And that is another way, now, of approaching the problems of the end of life at a time when the euthanasia of humanity is at issue as a fatal consequence of a shutdown, the *twilight of place* which no one is turning a hair at.

6 The Expectation Horizon

> Poetic creation is the creation of expectation.
>
> Paul Valéry

The feeling of insecurity that has crept up over the last dozen years or so in the city is not only linked to the discourteous acts of so-called 'incivility' currently plaguing city-dwellers. It is, it would seem, a symptom of a new expectation horizon, a third kind of horizon after 'revolution' and 'war', the Great War, the 'war to end wars'. I am talking about the expectation of the integral accident, this Great Accident that is not merely *ecological*. The latter has been part of our general mindset for the last thirty years or so. The integral accident is also, and above all, *eschatological*. It is the accident of a world now foreclosed in what is touted as 'globalization', this internationalization at once desired and dreaded, now the subject of endless debate, as though the anthropological horizon of ideas and ideals suddenly felt blocked off, both by the foreclosure represented by a geographical lockdown and by the suddenness of worldwide interactivity of exchange.

There is immense expectation now, in fact, of an integral accident. Horror films are not just the formatted products of the Hollywood dream machine and of the bid to systematically scare viewers or cinemagoers, as though hell lay at the heart of the world. No. What this is about is the recent emergence of an end-of-the-world feeling – an end in no way apocalyptic or millenarian, synonymous with some end of History but, more simply, an end of geography, as though the overhyped consumer society had finally consumed planetary space-time and been duly taken over in turn by the new communications society.

While ancient societies were almost all agoraphobic, shut in on themselves in their closed cities, within their outsized enclosures, postmodern societies suddenly seem claustrophobic, as though the open city of our day only leads, in the end, to exclusion.

'Completion is a limit,' Aristotle announced in his second axiom. The foreclosed world of economic and political globalization is effectively the ultimate limit of the geopolitics of nations, and the feeling of panicky insecurity felt by whole populations, along with the gigantic migration flows of the imminent repopulation of continents, are proof of this. The warning costs nothing and our democracies would do well to heed it before future tyrants use it to their own advantage.

And so, what is emerging, alongside the pollution of substances (of air, water, fauna and flora) at the very start of the twenty-first century, is the sudden pollution of distances and of the intervals that make up the very density of our daily reality; of this real space of our activities that the interactivity of the real time of instantaneous exchanges eradicates: the 'grey ecology' of the pollution of the life-size that rounds off the 'green ecology' of the pollution of nature by our chemicals and other products.

Here, we might mention the emergence of two currents of thought that are in no way antagonistic but complementary: *substantialism* (or, if you prefer, materialism) and *accidentalism* (or, if you prefer, spiritualism).

How can we fail to see that the primary political movement of the widespread accident is, of course, the one commonly referred to as 'the Greens'? A movement that is naturally more concerned with the pollution of material substances than with the pollution of time distances that reduces to nothing, or almost nothing, the extent and duration of our habitat; this human environment that, besides matter, yet possesses geophysical dimensions and proportions that are unextendable.

Proportions every bit as vital as water or the air we breathe for those who already fear that the great Locking Up of the seventeenth century (at the origins of the Revolution of the Enlightenment, according to Michel Foucault) might be reproduced – only, this time, not on the scale of the asylums or prisons of the Ancien Régime, but on a scale encompassing the entire world.

This is why we urgently need a second political movement focused on the integral accident that would complement the first – an eschatological party, this one, parallel to the ecological party officially recognized today.

Like the highs and lows of stereophony, this twin ecopolitical movement would create the effect of a field, a raised profile now indispensable to the right as well as the left in our democratic assemblies, since, as we all sense, this classic political representation will not be able to survive in the absence of a genuinely geopolitical definition of ecology. In other words, unless it takes account not only of the famous 'imperative of responsibility' of elected representatives, but also the 'precautionary principle' and the principle of vigilance of scientists and other decision-makers running the show.

In this sense the crisis or, rather, the accident in 'representative democracy' has nothing short-lived about it, since the televiewer-citizen can't be governed like an unregistered student or a nineteenth-century reader, his vision of the world being literally completely different. This is something that certain ecologists have just cottoned on to, such as the *Grünen* in Germany, who are now bent on coming up with a better interpretation of the notion of globalization itself – an ecological as much as an economic variety.[1]

'The civilized world should take seriously the growing threat of Terror on a catastrophic scale,' declared George W. Bush, on 15 March 2002. Seriously, certainly, but not tragically, otherwise we would end up lapsing into nihilism and leaping, without any transition, from the euphoria of the consumer society to the neurasthenia of a society of dereliction about which Karl Kraus was clearly apprehensive when he wrote, in 1914: 'Shrouded in the neurasthenia of hate, all is truth.'[2]

How can we fail to realize the extent to which, today, the game of politics has been defused and debased by this 'new idea' of so-called happiness carried by the looming shadow of the Revolution of the Enlightenment – but also by that of the Terror? How can we fail to see how powerless we are to tackle the major hazards and great disruptions looming large, in the face of which our hedonistic culture is basically defenceless?

Geopolitical ecology would also mean this: facing up to the unpredictable, to this Medusa of technical progress that literally exterminates the whole world.

Certain of the powers that be already reckon that the great gut-wrenching and divisive revisions of 'geopolitical economics' can't happen without some terrible worldwide crisis that, by terrifying each and every one of us, would shock nations and peoples into a sudden global stocktaking. But this is to forget a little too fast, it would seem, that fear is a bad counsellor, as all dictatorships have proved, since antiquity.

Since the last century each of us has learned from experience that dictatorships are not 'natural disasters'. They are created with the help of numerous unavowed collusions, in particular the collaboration of collective emotion. Who can forget the mass alarm about *Lebensraum* whipped up by the naturalist ideology of the Nazi movement?

Now let's turn to an event that is fairly minor but that says a lot about the ambient anxiety. In France, some little time ago, a National Union of Disaster Victims was set up, bringing together some sixty aid associations for the victims of accidents, ranging from the Abbeville floods to the Toulouse explosion via road accidents.

This national union now passes itself off as the sole negotiating partner confronting the authorities. A forerunner of some future eschatology party, this union of associations gives us a foretaste of the possible emergence not only of the coordination of 'victims' unions', but especially of the coordination of a party of 'casualties of life' that would replace the party, in the throes of extinction, of the exploited; those workers for whom socialism represented, once upon a time, the demand for justice.

But, here, the rampant ideology is not so much about a legitimate duty to protect populations; it is about a 'precautionary principle' taken to the absurd extreme of the myth of comprehensive insurance.[3]

'The idea of protection haunts and takes up the whole of life,' claimed one of the great exterminators of the twentieth century. But this paradoxical claim of Adolf Hitler forces us to go back over the origins of the various 'expectation horizons' that have preceded the one of the Great Accident of which ecology today presents as a symptom.

Since the eighteenth and nineteenth centuries, three types of expectation have, in fact, succeeded and overlapped each other, without a soul seemingly taking umbrage at the constantly escalating extremism they represent.

In the eighteenth century, it was firstly the revolution or, more precisely, revolutions, American and French, that were to lead to the suite of political upheavals we all know about, right up to the implosion of the Soviet Union at the end of the twentieth century, not forgetting the nihilist revolution of Nazism.

Buoyed by technoscientific progress, those political revolutions ushered in a whole host of industrial and energy revolutions, revolutions in transport and telecommunications, which we don't need to list here.

As Lenin explained, and he should know: 'Revolution is communism plus electricity.'

Parallel to this very first 'expectation horizon', the nineteenth century was to have a hand in generating the second, that of war, a Great War, whose geopolitical absurdity was flagged by the first worldwide conflict of 1914, following on from the Napoleonic epic. The other great conflict, the Second World War, was a total war, in which what was attacked at one and the same time was the human race as such, at Auschwitz, and its environment, at Hiroshima. This is to say nothing of the quarantine years of the balance of terror between East and West, that Third World War that remained undeclared under the pretext of 'nuclear deterrence' between the two antagonistic blocs. But the militarization of science and the arms race involving weapons of mass destruction that it gave rise to were soon to reveal just how atrocious this undeclared war was.

There is no need to spell out the strict correlation between these horizons of expectation, 'war' and 'revolution' mutually reinforcing each other in the name of a technical and political Progress that remains uncontested, except by a handful of heretical thinkers.

On this subject, let's hear it from one such heretic: 'In the nineteenth century, the notion of revolution–rebellion quickly ceased to represent the idea of violent reform, due to how bad things were, and instead became an expression for the overthrow of what exists as such, whatever it might be. The past having become the enemy, change for the sake of it has become what matters,' wrote Paul Valéry in the 1930s,[4] before rounding off this statement of the bleeding obvious with this: 'We are the greatest creatures of habit of all peoples, we French, who have turned revolution itself into a routine.'[5]

That is probably one of the unacknowledged causes of the defeat of France in 1940, even if the war of extermination had already long trumped the Revolution of the Enlightenment with the night and fog of the totalitarianisms.

Yet, over the course of those years that were so fatal for humanity, a few women glimpsed the truth of things more lucidly than many a statesman. Coming after Simone Weil and Hannah Arendt, Brigitte Friang tells us about the period between the wars:

> All through my childhood, I heard talk of war [. . .]. Films, Verdun, 'a glimpse of history' whose cannonfire haunted my little girl's nightmares. Henri de Bournazel and Commandant Raynal were as familiar to me as Bibi Fricotin or Zig and Puce! This kind of intimate company is rarely inconsequential. War, war! That was the key word, the definitive word, the leitmotiv. It was so infallible, it didn't fail.[6]

And that about wraps it up, you'd think, but no, there is more, a wisecrack from Pierre Mendès-France, delivered in 1968: 'It's 1788, only without the revolution of the following year.' And that, indeed, was the case. The events of that spring remained 'events', a sort of literary Commune, and nothing more. The concept of 'revolution' had exhausted its ideological fecundity and after that remained only a mute anxiety, the expectation of some nameless catastrophe in which fledgling ecology was shortly to take over from the Big Night that was to end, what's more, in the implosion of the USSR soon after Chernobyl, a premonitory cataclysm of a future not so much radiant as radioactive.

And that is how the twentieth century came to a close, after more than two hundred nasty wars yielding hundreds of millions of victims: the First World War, 15 million dead; the Spanish Civil War, 500,000; the Second World War, 50 million; the Korean War, 4 million; the Iran–Iraq war, 500,000. As for the second Gulf War, there is talk of 200,000 victims. And it is not over yet, it seems . . . .

But to stick to revolution and war for a moment, let's turn our gaze towards what is happening, giving discernment a chance. Beyond ethics, it seems that bioethics is troubled these days about the major risks that the 'revolutionary' discoveries of the biotechnologies are making the human race run, and which will in the near future lead to the threat of a sort of cellular Hiroshima in which the genetic bomb will, this time, ravage man's very form, just as the atomic bomb, in its day, shattered the horizon of man's environment.

As far as that goes, there is no lack of threats to life, between medically assisted procreation, human cloning and, now, the right to assisted death and euthanasia, not to mention biological weapons. Everything is in place for the Great Accident in the Book of Life.

At the start of the year 2002, for example, as though to symbolically mark the dawn of the third millennium, Dr Severino Antinori, whose gynaecology clinic is in Rome, in the Eternal City, announced, like the archangel Gabriel, the imminent birth of the first human clone. Certain anonymous and carefully concealed female gene donors were, apparently, getting ready to give birth thanks to the procedure known as 'reproductive cloning'.

And so, the hope of an eternity of the soul and the sun of the resurrected are having to compete with the shadows of the retorts and stills of an evil genetic genie. The resurrection of the dead has been swapped for the duplication of the living. And, suddenly, the good 'Miracle Doctor' was proclaiming that there was no doubt that between December 2002 and January 2003 a cloned child would be born, the first 'replicant' of the human race. Why not at Christmas?

By way of conclusion, let's go back now to this 'feeling of insecurity' that has come over the masses today and that already largely conditions the political life of Western nations.

Despite the threat of an unemployment that is structural and definitive for certain categories of people hard hit by the boom in automation of postindustrial production, the anguish now clearly palpable does not seem to be linked to such exclusion from employment, nor to the 'incivility' plague or domestic violence either, but, more profoundly still, to anguish over the failure, also definitive, of the Progress in knowledge that until this moment so strongly marked the age of industrialization.

In fact, the very first expectation of 'revolution' went hand in glove with the expectation of a progress at once philosophical and scientific that was itself to be swept aside by the hurricane of war; of a total war of which the militarization of national economies, over the course of the twentieth century, already flagged the devastating magnitude. The only thing it allowed to survive in people's consciousness was this feeling of fear – and often of hate – that today marks societies of abundance.

On this score, over to Karl Kraus once more: 'Ever since humanity bowed to the economy, all it has left is the freedom of hostility.'[7]

In 1914, the date of this premonitory phrase, it was still only a matter of a deadly rough draft of a new 'war economy' that was to bring down the nations of Europe alone. But in these early days of the twenty-first century, which is our century, it is a matter of the conclusion of this political economy of disaster.

From now on, as every one of us senses, fears and dreads, the world is closed, foreclosed, and ecology has suddenly become the third dimension of politics, if not its very profile.

After the city-state and the nation-state, the outsize federation of the European Community and other groups like it is merely the pathetic mask of a geopolitical bankruptcy that goes by the assumed name of globalization – an integral accident in a political economy that has just reached the geophysical limit of its field of action.

7 Unknown Quantity

'Luck is like us,' George Bernanos once wrote. Indeed, if once upon a time life was still a theatre, a stage with its transforming sets, daily life has now become sheer luck, a never-ending accident, with its many new developments, the spectacle of which is inflicted on us at every moment via our screens.

Actually, the accident has suddenly become habitable to the detriment of the substance of the shared world. This is what the 'integral accident' is, this accident that integrates us globally and sometimes even disintegrates us physically.

And so in a world from now on foreclosed, where everything is explained by mathematics or psychoanalysis, the accident is what remains unexpected, truly surprising, the unknown quantity of a planetary habitat totally uncovered, overexposed to the eyes of all, from which the 'exotic' has suddenly disappeared to the advantage of the 'endotic' championed by Victor Hugo when he explained to us that 'it is inside yourself that you should look at the outside'[1] – a terrible admission of asphyxiation, if ever there was one.

'The ego is originally all-inclusive, but later it separates off an external world from itself. Our present sense of self is thus only a shrunken residue of a far more comprehensive, indeed all-embracing feeling, which corresponded to a more intimate bond between the ego and the world around it.'[2]

Originally Freud was perhaps right, but, in the end – and that's where we are ecologically – when our feeling embraces all once more due to the fact of the temporal compression of sensations, we'd better watch out, for this will then be the great reduced, incarceration in the tiny cubby hole of a once 'oceanic' feeling for the world, suddenly reduced to claustrophobic suffocation.

This, in any case, is what is bitterly admitted by astrophysics: 'The rupture with the whole slew of great cosmic events is one of the causes of the malfunctioning of human societies.'[3]

For proof of this astronomical fracture caused by globalization, let's now look at a phenomenon of eccentric pollution, brought to light (and how!) by a society for the protection of the night sky.

Because of the scale of light pollution caused by overpowerful electric lighting, two thirds of humanity are now deprived of true night.

On the European continent, for instance, half the population is no longer able to see the Milky Way, and only deserted regions of our planet are still really plunged into darkness at night. This has reached the point where it is no longer only the night sky that is threatened but indeed the night itself, the great night of interstellar space; that other unknown quantity that, nonetheless, constitutes our only window on the cosmos.[4] The situation is such, furthermore, that the International Dark-Sky Association has just launched a surrealist petition to get the night listed on the world heritage list as a heritage of humanity!

'The World is deeper than the Day thinks,' wrote Nietzsche, while it was still a question of sunlight. But already, here and there, and often everywhere at once, contemplation of a screen not only replaces contemplation of script, the written word, the writing of history, alone, but also contemplation of the stars. So much so that the audiovisual continuum has superseded the – substantial – continuum of astronomy.

In this 'disaster writing' of space-time, where the world becomes accessible in real time, humanity is struck with myopia, reduced to the sudden foreclosure of a seclusion triggered by the accident in time of instantaneous telecommunications.

From that moment, to inhabit the integral accident of globalization is to block, to choke off not only the view, as Abel Gance hoped, followed by the filmmaker apostles of cinemascope, but also the daily life of a species that is nonetheless endowed with the motion of being.

At this stage of incarceration, terminal history becomes a *huis clos*, a hearing in camera, as camp detainees so rightly put it: 'Our horror, our stupor, is our lucidity.'[5]

Everything is there, already there, already seen and soon, even, already said. All that's left after that is to wait the long wait for a catastrophic horizon that outstrips the geographic horizon of the rotundity of the earthly star.

And so, the local accident located here or there is trumped by the great accident, the global accident that integrates, one by one, the whole set of minor incidents along the way that once characterized societal life. This 'great lockdown' then puts an end to banishment, only to promote a sequence that is causal, this time, since, from now on, 'everything arrives without the need to leave', to go towards the other, the distinctly other, as we once went towards a landscape's horizontal limit in days gone by.

Here, and whether Nietzsche likes it or not, it is no longer God, the Father, who dies, it is the Earth, Mother of the living since the beginning of time. With light, the speed of light, matter is being exterminated. The telluric accident of the earthquake is succeeded by the seism of a timequake involving this worldwide time that erases all distance.

In this abrupt telescoping of successive events that have become simultaneous, it is expanse and duration that are erased.

After having been disintegrated by means of the nuclear bomb, matter is now being exterminated by means of acceleration, the specular bomb of screens, those mirrors of time that cancel out the horizon.

Within the enclosure of its terrestrial environment, reaching the threshold of an interstellar void that, far from having conquered, humanity dreads, 'The ultimate experience is an experience of what is "outside everything", when that everything excludes everything outside.'[6]

At the point we are coming to in the twenty-first century, what is looming is therefore not so much the end of history as the end of multiple times. Suddenly, with the extermination of the distances of the local time of geophysics, faced with the light years of a purely astrophysical time, 'man has in a way joined the omega point, which means there is nothing other than man any more and there is no outside any more outside him.'[7]

Here is the ultimate figure of philofolly, that is to say, of the accident in knowledge whereby 'man affirms all by his very existence, embraces all including himself within the closed circle of knowledge.'[8]

Then, within the limits of this closure, something outrageous lies in wait, not as in 'the exile of madness' experienced by the deviants locked up in the asylums of the nineteenth century any more, but in the exodus of the philofolly of the high and mighty; those mad scientists once stigmatized by Swift, rendered powerless by the maniacal outrageousness of discoveries that aren't so much superhuman as fundamentally inhuman.

How else, other than as a major clinical symptom, can we interpret the fact that more than ten million people in France have become hooked on video games, frequenting networked gaming rooms the same way a person would go into an opium den, logging on to the Internet the same way you would get yourself a fix?

A panic phenomenon of dependence, the vogue in 'on-line games' has given a new dimension to what psychiatry used to call a loss of the sense of reality, driving adults and adolescents into a groundless parallel world, where each individual gradually gets used to inhabiting the accident of an audiovisual continuum, independent of the real space of their life.

At this stage of cybernetic seclusion, presented as the crowning achievement of Progress, where the most trifling bit of information and the most trivial event zip around the world in an instant, globalization puts paid to 'revolution' just as it does to the classic 'world war'. For, thanks to the ubiquity of television, the slightest incident can become 'revolutionary' and the most piddling attack relayed on a loop can take on the gigantic proportions of a worldwide conflagration!

That is finally the effect of this omega point humanity has reached, a 'meteorological' effect that reproduces the one where a butterfly beats its wings in an Amazonian rainforest and causes a hurricane in Europe – just as the El Niño phenomenon is now playing havoc with the climate of the globe.

In this sense, as Maurice Blanchot pointed out in relation to the Age of Enlightenment, 'Shutting in the outside means setting it up as an interiority of expectation or exception; this is the requirement that leads society to cause madness to exist, meaning to make it possible.'[9]

This is precisely what is happening to our globalized societies, where the local is the exterior, and the global the interior of a finite world, exclusively defined by the existence of networks of instantaneous information and communication, to the detriment of any geopolitics, since the real time of (economic, political) exchanges wins hands down over the real space of the geophysics of the world's regions.

By accelerating, globalization turns reality inside out like a glove. From now on, your nearest and dearest is a stranger and the exotic, a neighbour. The deregulation of transportation is topped by the derangement of a foreclosure that triggers exclusion of the 'close' to the momentary advantage of any 'far-off' whatever that you happen to stumble across in the telescoping of civilizations.

The expectation horizons of a past three centuries old that is now over – those of total revolution and total war – have been outpaced by the anguished expectation of the (eco-eschatological) Great Accident of which industrial accidents and terrorist attacks are only ever prefigurations, symptoms of a complete reversal in the orientation of humanity.

But this very latest attack is inseparable from the accident in time,[10] since the acquisition of the speed of light shatters the plurality of social times and favours a generalized synchronization of action, interactivity then outpacing customary activity. Teleaction that eliminates not only the long durations of familial and social relations, but also those of the political economics of nations in tandem with their military strategy.

Whence the recent drastic overhaul of the substantial war (Clausewitzian, if you like), boosting this anonymous and fundamentally risky accidental war, that hooks up declared hostilities to industrial or other accidents, thereby promoting a fatal confusion between attack and accident.

Global terrorism is, in fact, like fate and its 'strokes of luck', good or bad, the force of destiny completing the force of the traditional army equipped with weapons of mass destruction, inherited from the age of world war, now over.

But listen to Victor Hugo: 'I have defined and delimited the "state of siege": if anarchy is the arbitrary in the street, the arbitrary is the anarchy of power.'[11] From now on, the 'state of siege' is globalization, this foreclosure that transforms, or soon will transform, every state into a police state, every army into a police force and every community into a ghetto . . . .

And so, globalization's closed-field effect is nothing less than the progressive strangulation of the legitimate state of representative democracy, the society of strict supervision taking over from the society of local seclusion. After the standardization ushered in by the industrial revolution, synchronization (of opinions, of decisions) has come to set up an ultimate model of tyranny: the tyranny of this real time of forced interaction that replaces the real space of action and its free reaction within the expanse of a world that is open . . . but only for a little while longer.

If interactivity is to information what radioactivity is to energy – a contaminating and disintegrating capability – then the integral accident in time causes conflicts in the *socius* and its intelligibility to accumulate, making the whole world opaque little by little. After the accident in substances, meaning matter, the time of the accident in knowledge is upon us: this is what the so-called information revolution really is and what cybernetics really is: the arbitrariness of anarchy in the power of nations, the different powers of a community not only thrown out of work by automation but further thrown out of whack by the sudden synchronization of human activities.

Part II

8 Public Emotion

According to Clausewitz, 'war first surfaces in the art of holding a siege.' This military art is thus opposed to the tumults of the origins of the history of conflicts.

Today, as everyone can see for themselves, 'hyperwar' resurfaces in the art of provoking panic, thanks to the tools of mass communication.

A purely media phenomenon, this situation in turn entails reinterpretation of the classic notion of deterrence. 'Military' deterrence in the recent past, 'civil' deterrence in the near future or shortly after: the threats to democracy are numerous.

In fact, it is definitely the fortification that, in history, has best embodied the desire for deterrence of the different powers. Isolated, linear or strung out in a network of strongholds, the rampart signals a desire for deterrence in relation to some massive aggression, but as Thierry Wideman rightly points out, 'From the point of view of strategic thinking, a global theory of conventional deterrence seems unworkable, unless it is based on Clausewitz's theory of the superiority of the defensive raised to the status of an axiom, the multiplicity of variables effectively making any generalization impossible.'[1]

As an operational strategic concept, deterrence only made sense with the advent of nuclear power.

Military deterrence in the recent past, still useful, apparently, in the face of certain 'rogue states'. Civil deterrence in the near future in the face of the threats of a latent, if not patent, hyperterrorism . . . In writing this, it is vital to specify that at its inception war was already part and parcel of that collective mass psychosis that afflicts the besieged, buried alive as they are behind their protective enclosures.

Siege warfare was at the political origins of the history of cities and of nations. Mobile warfare came later on. We should also note that territorial conflicts have never stopped accelerating, finally turning into this 'Time War' that will soon overrule the war of the geostrategic space of empires, and in which sea power will be taken over by the power of air-naval and, finally, air-orbital forces.

Today, though, globalization and its poliorcetic foreclosure are spreading on a planetary scale. But by the same token, what is surfacing with this global state of siege are no longer the enclosure and its colossal fortifications, despite the illusory anti-missile system of the United States. First and foremost what it produces is the inordinate spread of panic, a panic that is still mute, certainly, but that never ceases to grow at the rate of all the accidents and disasters and 'mass terrorist attacks' that point to the emergence, not so much of some hyperterrorism as of this post-Clausewitzian 'hyperwar' that outpaces all political givens regarding conflicts, national or international.

Damaging strategy, in other words, the geostrategy that so long rejected its new chronostrategic dimension, this sudden internationalization of real time imposes *ex abrupto*, a different tyranny – that of instantaneity and ubiquity – not only on military commanders and planners but on the democratically elected politicians who are supposed to be running the show.

In fact, after mass and energy (atomic or otherwise), war now opens into its third dimension: information that is instantaneous, or as good as.

Whence the untimely emergence of information astrostrategy that each and every one of us benefits from, whether military or civilian, soldier-citizen or terrorist, not to mention simple common-law criminal.

At this point in time, as the third millennium kicks off, what is dawning in people's mentalities is what some like to describe, euphemistically, as a feeling of insecurity. And this is nothing but the symptom of mass panic of the besieged targeting, in the first instance, the metropolitan concentrations, veritable 'resonance chambers' that they are, of a type of population movement no one really regulates.

Actually, the more the contemporary city-dweller is subject to diffuse and uncertain threats, the more he or she tends to make political demands for someone at the helm to be punished, for want of an avowed guilty party. This is what the clandestine terrorist takes advantage of, thereby directly threatening the representative democracy of assemblies and even, lately, the democracy of opinion created by the major media outlets, thereby boosting a democracy of public emotion that is nothing less than the poisoned fruit of the panic phenomenon referred to above.

In fact, what emerges alongside the necessary formation of public opinion by the sundry media outlets is the unheard of possibility of a public emotion whose unanimity would be merely the symptom of the decline of any true 'democracy'. And this would in turn pave the way for a conditioned reflex, no longer 'psychological' but 'sociological', a fruit of the panic-ridden terror of populations faced with the outrageousness of the broadcasting of real or simulated threats.

And so, after the launch, almost a century ago, of ecstatic consumption, we would look on powerless, or very nearly, at the booming of a form of communication no longer 'ecstatic' but openly hysterical. Audiovisual interactivity has already mastered the secret of this, thanks to the possibilities of an instantaneous commutation in collective emotions, and this, on a worldwide scale, the synchronization of mindsets cleverly rounding off the old standardization of opinions of the industrial era.

With mass terrorism, this hyperwar in which mass no longer bears reference to armies and armoured divisions but to civilian victims, the unarmed populations have become the exclusive parade ground, the ground that takes over from the battleground of the military campaigns of yore.

In this war on civilians that borrows a number of features from age-old civil wars, it is still the war of movement that carries on, speeded up, with its tactics and its tricks. But the 'movement' now means, above all, the panic-stricken flows of terrorized populations.

And so the serial killer of 'organized crime' is outclassed by the mass killer of 'organized terrorism' in the age of the imbalance of domestic terrors.

Lumped together in the metropolises, urban populations suddenly become the breadth and depth, but especially the height, of the action engaged, resulting in the now emblematic dimension to the VHB, or very high building, since the collapse of the Twin Towers.

But what subsists of the concrete and down-to-earth in this strategy of hypertension is demo-topographic concentration, not as once upon a time within the fortified market town, nor even the enclosure of independent cities, but now within the megapolitan nebula harbouring tens of millions of inhabitants.

At the end of the day, this is the metropolitics of terror that is gearing up to resurrect the geopolitics of size, national or imperial. Everywhere you look now, the scale of terror dominates the scope of space, the real space of nations and their old common borders.

Communicable at a distance and in real time, panic flows have once and for all replaced the old tactical movements of military units of days gone by. It is all too easy, in fact, to imagine the day when an 'accident' (telluric or otherwise) or pollution incident (maritime or other) will set off regime change in the targeted nation, along the lines of what happened in Spain after the Madrid terrorist attack.

Once, not so long ago, the French monarchy dared not imagine the worst and so it endured the terror of the Revolution, followed swiftly by the Empire. Shortly, if we are not careful, the same could happen to democracy in Europe. At any moment, a transpolitical disaster could cause us to relive the death pangs of bygone political revolutions, to the very great detriment of public freedoms.[2]

But let's get back to this tele-objective panic produced by the telephoto lens and the various mass movements it gives rise to.

Note, for instance, the blitzkrieg waged by the German armoured divisions that, in 1940, propelled onto the roads of France some 12 million civilians terrorized by the Fifth Column. Strangely, this figure is exactly the same as the number of Spanish citizens who marched in the streets of Iberian towns after the terrorist attack on Atocha railway station on 11 March 2004. These same citizens were, the very next day, to overthrow the Aznar government, against all the forecasters' predictions.

In this other brand of blitzkrieg, panic is the main force of organized terrorism and it is no longer so much the discipline of the troops as the lack of discipline of the hordes that becomes decisive.

This is the impetus behind the strategic programming of terrorist attacks, either for the nightly newscasts, as in Paris, fifteen years ago, or for the day before elections, as in Madrid in 2004, thereby provoking a public emotion that was to shatter the indispensable serenity of the democratic vote, along with the opinion of future electors.

Faced with this psycho-sociological condition of the horrified masses, the old 'science of the defence and attack of strongholds', or poliorcetics, is transformed.

The very last bastion of public freedoms is merely the mass of potential victims!

By way of example, we might recall Mao Zedong's China where the United States was only ever seen as a 'paper tiger' since, with over a billion inhabitants behind it, Communist

China was not afraid of a nuclear war that would involve hundreds of millions of dead.

Public opinion or public emotion? This domain finds itself in the same situation as so many others where the community of interests conducive to political action gives way to a 'community of emotion' open to all kinds of manipulation.

And so, a new Anglo-Saxon practice known as storytelling now sees professional storytellers intervening in corporate life. Their job is to tell stories to wage-earners in order to foster certain behaviours and certain emotions in them, within the framework of job restructuring or relocation, an intervention that clearly flags the new importance of emotion management in business administration.[3]

But we must not confuse the feelings we might experience and the emotions we might feel, for feelings can be submitted to the test of reason, thereby avoiding any untoward reaction, whereas, emotion, on the contrary, easily escapes all control in mob phenomena.

Since the age of revolutions, this type of mob rule has constantly overturned the very form of the 'republic' and, consequently, of our democracies. You only have to look at the 'rape of the crowds' by the different totalitarian regimes in the course of the twentieth century.

Public opinion is supposed to be built up through shared reflection, thanks to the freedom of the press but, equally, to the publishing of critical works. Public emotion, on the contrary, is triggered by reflex with impunity wherever the image holds sway over the word. Easy to trigger through any over-the-top *mise en scène*, the herd effect of whipping up collective emotion meshes perfectly with televisual cinematics, as well as with the interactivity of cybernetic technics, madly stoking every kind of frenzy.

Whereas republican opinion rested, from the very beginning, on the art of oratory and reading, post-republican emotion rests, for its part, on sound and light. In other words, on the audio-visibility of a spectacle or, rather, of an incantatory liturgy that is only apparently secular . . . witness the characteristic abuse of rebroadcasting not only of commercials but equally of terrorizing events.

In its grandiloquent fashion, this media phenomenon overrides the state itself in a sort of accident in political substance that has particularly far-reaching consequences for the future of republican freedoms.

'The electorate no longer knows the party!' wailed a German journalist after the stinging swing in the French elections of March 2004. He went on to specify that 'the actions of those who govern the countries of Europe today are undermined by fear (of economic stagnation, of unemployment, etc); in other words, fear of the future.' Strangely, this list does not even mention terrorism and its devastating effects on the Spanish government, kicked out only a fortnight earlier.

Here again, the electoral accident argument wins out over the argument of the terrorist attack, as in the explosion in the Toulouse fertilizer factory in September 2001, just a few days after the attacks on New York and Washington.

On the subject of the Toulouse ammonium nitrate explosion, we might hazard a gratuitous hypothesis: supposing that the reverse happened and that the Toulouse investigators had opted for the terrorist attack line of inquiry. Not only would French diplomacy have been quite different, but the Franco-German 'peace camp' would have gone up in smoke under the pressure of a public opinion traumatized by the scale of the disaster – as happened later in Madrid.

Suppose now that the new investigations under way in Toulouse ended, shortly, in flagrant proof that the Toulouse tragedy, in which over 30 people died and over 2000 were injured, was indeed the result of a twin attack: 11 September in the United States and 21 September in France, at Toulouse, the city of the European aerospace industry, host to Aérospatiale.

In that case, what would be the consequences of this situation with its staggered front on European geopolitics, as well as on the fate of a French president suspected of having covered up a mass terrorist attack, while the president of the United States only lied about the existence of weapons of mass destruction? But this is, of course, just a simple hypothesis of politics-fiction.

'Fear has been the ruling passion of my life,' Roland Barthes confessed, before being wiped out in a road accident. I very much fear that, tomorrow, this individual passion will be the collective passion of societies crippled by the untimely nature of catastrophic events whose repetition winds up fostering fatalism, at least, if not despair.

When the unexpected is repeated at more or less constant intervals, you come to expect it, and this 'expectation horizon' then becomes an obsession, a collective psychosis open to every kind of manipulation, every kind of destabilization of public order.

Of course, with mass terrorism and its instant impact, fear can't stay private and restricted to the minority for long. It tends inevitably to become public and available to the majority, with the consequence that any kind of true courage is not possible, unlike with individual fear, but only the indifference that preludes the silence of the lambs.

From that moment, we can guess where the administration of public fear finally leads: to this civil deterrence that not only succeeds the military deterrence of the Cold War era, but especially the 'fear of the policeman' of the policed societies of days gone by. There is one difference, though, since such an administration will no longer be 'republican' but will be entirely bound up with the mass media.

In the face of the hyperwar, Clausewitz's theory of the superiority of the defensive over the offensive is outmoded by the very nature of information – this third dimension of conflict after mass and energy. So it is urgent that we study the question, at once psycho-political and socio-strategic, of domestic terror.

The first indication of such anticipation in public emotion is provided to us by the American concept of a 'war of zero deaths', with its surgical strikes and its preventive wars that spare the soldiers fighting, at least, if not civilians.

We know what flows on from such fibs. A paramilitary concept, the preemptive war as a matter of fact signals a strategic grand illusion: the one where the offensive is no longer anything more than a disguised defensive against an asymmetrical adversary who is disqualified as a fully-fledged 'partner' in some wargame where the classic alternation between attack and counterattack is blocked, on the one hand, by an enemy who refuses to do battle, and, on the other, by development of an electronic arsenal of lures and techniques for avoiding any real engagement.

As an age-old proverb has it: 'Fear is the worst of killers; it doesn't kill you, it stops you from living.' It even stops soldiers from making war, according to the rules of political propriety!

This has nothing to do with the threat along a state's borders or the assault of some invader or other. The phenomenal migration flows of dire poverty or of mass tourism have long replaced these. And that is to say nothing of the pending shift, also massive, of well-off populations, newly dissatisfied with the lack of comfort and security offered by the great metropolises of affluent countries.

No, the administration of domestic terror imposed by the various major hazards has absolutely nothing to do with the threats of a recent past. The equation is radically different and 'armchair strategists' would do well to think twice before engaging in military responses to terrorism, responses that are, in the end, nothing but tragic 'distractions'.

After the Cold War and its apocalyptic threats of annihilation comes the time of this cold panic of an organized terrorism likely to inflict analogous disasters.

Imperceptibly, with the decline of the nation-state, we are seeing the end of the monopoly on public violence enjoyed by the state, triggering the ascendancy of a privatization of domestic terror that not only threatens democracy but the legally constituted state.

Europe, today inordinately enlarged, can't go on for long turning a blind eye to these issues that are not so much political any more as 'metropolitical', since the demographic concentration of its populations in megalopolises has gradually shifted the old theatre of operations from the country to the city, with the 'carpet bombing' of the mid-twentieth century prefiguring the 'mass suicide bombing' against densely-populated urban agglomerations at the very beginning of the twenty-first century.

And so, the very notion of defence is radically transformed. After the military defence of nations and the civil defence of urban populations it seems that there is an urgent need for a new line of inquiry.

On top of national security, based on the armed forces, and social security, underdeveloped as it is in a number of democratic states, we must now add the crucial issue of human security, which would extend the old public interest of the state.

As the former United Nations High Commissioner for Refugees, Mrs Sadako Ogata, recently declared: 'September 11, 2001, demonstrated that no state, not even the strongest militarily, is capable of protecting its citizens any more, not even within its own borders.'[4]

Faced with this alarming assessment, which introduces the temptation of a sort of nihilism, not only in defence (as happened in certain Nordic countries before the Second World War), but in the public arena, with the city as epicentre, it might perhaps be useful to take a closer look at the historic shift in the armed forces. This shift has taken us, as we have seen, from siege warfare, with the domination of weapons of obstruction (ramparts, fortifications of all kinds), to the war of movement and, finally, to this *blitzkrieg*, or lightning war, in which weapons of destruction supplanted urban and other entrenchments until the days of the deterrence strategy. With the latter, which introduced the 'non-battle,' the relative inertia of the balance of terror greatly favoured not only the arms race and eccentric proliferation of weapons, but, more especially, the development of those 'weapons of mass communications' that are today throwing the old geopolitics of nations into turmoil every bit as much as the stability of a military culture in disarray for over a decade – in other words, since the fall of the rampart of the Berlin Wall and the collapse of the keep of the World Trade Center in New York.

The current latent conflict simmering away in the United States between the State Department and the Pentagon is a fatal sign of this panic. But so is the US Army's split personality project, that will pit the historic army against a second army now on the drawing board, an 'anti-panic' army, designed to mop up the damage done to the legitimate state and to do so in a public arena undergoing accelerated privatization.

Whether we like it or not, public space and public authority are indissociable and any attempt to split them is equivalent, sooner or later, not only to undermining national security but especially to undermining human security, with the obvious risks of genocide that this entails.

The much-touted 'precautionary principle' of the ecologists thus applies above all to this necessary stability of public law and its real space, as the setting of any democracy.

We should point out once again that this new notion of 'human security' that was recently adopted in Canada and Japan – a probable result in Japan of the major earthquake at

Kobe and the sarin gas terrorist attack in the Tokyo underground – could well contribute to the outlawing of this *uncivil* war that threatens to wreak havoc, in the near future, not only on the legitimate state but indeed on the whole panoply of civilizations.

After the privatization of energy, the privatization of the public arena will inevitably lead, not to the professionalization of the public (police) force now, but to military anarchy. This veritable 'defence nihilism' will no longer involve an openly declared enemy – much as the Swedish movement Forvarsnihilism hoped, when it asked, in the 1920s, 'Is the invasion of our territory by another civilized people such a serious thing?'[5] It will impact on the military institution itself, as the basis of that 'right to defence' that subtends all political rights.

What can you say today, in fact, of the territorial invasion by a 'civilized people' when it is precisely a matter of mass terrorism using the complete array of the democratic amenities of transport and telecommunications provided by societies open to the most incredibly diverse exchanges of the age of planetary globalization.

CLAUSTROPOLIS or COSMOPOLIS? A society of enforced seclusion, as once upon a time, or a society of forcible control? Actually, the dilemma itself seems illusory, with the temporal compression of instantaneity and the ubiquity of the age of the information revolution. This interactive society is one in which real time overrules the real space of geostrategy, promoting a 'metrostrategy' in which the city is less the centre of a territory, a 'national' space, than the centre of time, of this global and astronomical time that makes every city the resonating chamber of the most incredibly diverse events (breakdowns, major accidents, terrorist outrages, etc). Break up of a social order will be triggered by the extreme emotional fragility of an aberrant demographic polarization, with megalopolises that will shortly bring together, not millions but tens of millions of inhabitants in very high towers where they will be interconnected in a network, and where the standardization of the industrial age will make way for this synchronization of collective emotion likely to do away with all democratic representation, all institutions, promoting instead a hysteria, a chaos of which certain continents are already the bloody theatre.

We should point out further that if interactivity is to information what radioactivity is to energy, deterrence is transformed: military deterrence or civil deterrence? That is the question!

And so it is no longer a matter here of going beyond the geopolitics of nations, nor of going back to the ancient poliorcetics of city-states. It is a matter of truly taking it to the limit: the mounting extremism of a hyperviolence that Clausewitz could never have imagined.

9 The Original Accident

According to Albert Einstein, events do not happen, they are there and we merely encounter them in passing, in an eternal present; there are no minor incidents on the way, history is merely one long chain reaction. Hiroshima, Nagasaki, Harrisburg, Chernobyl – simply instances of momentary inertia, the radioactivity of a place being analogous to the relativity of an instant.

Fusion, fission: the measure of power is no longer so much matter, but immateriality, energy output.

From now on, motion commands the event. After the 'earth worship' of the original paganism comes the terror worship of the original accident; this terror that is only ever a product of the laws of motion, as Hannah Arendt used to say.

In fact, it is urgent that we go back on the philosophical tradition according to which the accident is relative and contingent and the substance absolute and essential. From the Latin *accidens*, the word 'accident' signifies what arises unexpectedly – in a device, or system or product; the unexpected, the surprise of failure or destruction. As though this 'temporary failure' was not itself programmed, in a way, when the product was first put to use.

Actually, the arrogant primacy accorded to the production mode really does seem to have contributed to obscuring the old production mode/destruction mode dialectic (rather than simply consumption mode) in force in pre-industrial societies. Since the production of any 'substance' whatever is instantaneously production of a typical 'accident', then a breakdown or failure would not mean deregulation in production so much as the production of a specific fault – in other words, partial or total destruction. Fundamentally modifying research and development accordingly, we could then imagine some long-term planning of the accident.

Since this latter is innovated in the instant of scientific or technological discovery, perhaps we could turn things around and directly invent the 'accident' in order to then determine, afterwards, the nature of the famous 'substance' of the product or device implictly discovered, thereby averting the development of certain supposedly accidental catastrophes?

This reversal in perspective of the original accident, which vaguely smacks of mythology or cosmogonic theories like the big bang, really does seem to be the same as that operating in the 'dialectic of war,' that of the sword and the armour. It is, in other words, the perspective that shot to the fore with the strategic emergence of the 'war machine', in the immediate area around the ramparts of the citadel-state of Ancient Greece, which saw a contemporaneous political innovation, poliorcetics, the science of offensive and defensive siege warfare on fortified cities. Poliorcetics was to be the very origins of the future development of the art of war, that is, of the evolution of the production of mass destruction, throughout the ages, but most especially throughout the progress in weapons technologies.

The scientific and industrial production machine is doubtless merely an avatar or, as they say, blowback from development of the tools of destruction, from this absolute accident that is war, from this conflict pursued in all societies over the centuries, this 'great war of time' that never ceases to flare up out of the blue, here and there, despite the evolution in customs, the means of production and 'civilizations'. Its intensity never ceases to grow, either, with technological innovations, to the point where the latest energy, nuclear energy, at first appears as a weapon, at once armament and absolute accident in history.

The positivist euphoria of the nineteenth and twentieth centuries, this 'great movement of progress', would surely have to be one of the most insidious features of the bourgeois illusion aimed at covering up the fearful progression, as much industrial as military, in the mode of scientific destruction.

And, more precisely still, aimed at concealing the philosophical and political reversal of this absolute accident now making all substance, whether natural or manufactured, contingent.

'In the twentieth century we learnt the atomic nature of the entire material world. In the twenty-first, the challenge will be to understand the arena itself, to probe the deepest nature of space and time,' writes the British astrophysicist and Astronomer Royal, Sir Martin Rees.[1]

A little further on, extending this observation about the 'unknown quantity', Rees adds: 'More than fifty years ago, the great logician Kurt Gödel invented a bizarre hypothetical universe, consistent with Einstein's theory, that allowed "time loops", in which events in the future "cause" events in the past that then "cause" their own causes, introducing a lot of weirdness to the world but no contradictions' (p. 149).

By way of concluding these transhistoric words, Rees specifies further:

> A unified theory may reveal some unsuspected things, either on tiny scales, or by explaining some mysteries of our expanding universe. Perhaps some novel form of energy latent in space can be usefully extracted; an understanding of extra dimensions could give substance to the concept of time travel. Such a theory will also tell us what kinds of extreme experiments, if any, could trigger catastrophe. (pp. 150–1)

This would be a cosmic calamity and not just a terrestrial one 'in which the concentrated energy created when particles crash together could trigger a "phase transition" that would rip the fabric of space itself.'

According to the official astronomer to the Royal British Court:

> The boundary of the new-style vacuum would spread like an expanding bubble. In that bubble atoms could not exist: it would be curtains for us, for Earth, and indeed for the wider cosmos; eventually, the entire galaxy, and beyond, would be engulfed. And we would never see this disaster coming. The 'bubble' of new vacuum advances as fast as light, and so no signal could forewarn us of our fate.[2]

With this fantastic illustration of the dromosphere of the speed of light in a vacuum, we are at least in time to question the witnesses, those of Chernobyl, for instance, for in 1986 the time of the accident suddenly became for them, and finally for all of us, the 'accident in time'.[3]

Indeed, if the atmospheric currents at that period drove the contaminated clouds towards the west of the continent, the winds of history, for their part, drove its pollution towards the future, the setting sun of time.

And so, the past of the 1980s as a decade is intact, out of reach of the fallout from Chernobyl. But the future, on the other hand, is wholly polluted by the very long haul of nuclear radiation. If 'nature' is affected here and now, starting from that fateful day, it is the 'life-size nature' of future times that is already contaminated by the radionuclides of the year 1986.

The accident that occurred at the power station that day was well and truly an original accident. To prevent its becoming, tomorrow or shortly after, eternal, we are going to have to swiftly protect the area around the present against the future, as they once protected the area around the fortified city against the barbarians.

In the year 2007, the Chernobyl power station will be covered with the biggest concrete cape ever poured. The lifespan of this hopefully impermeable seal? One hundred years.

Actually, concrete quite simply turns into the time barrier and the resistance of a material (reinforced concrete) is used to ward off the disaster of Progress.

As colossal as the wall of a bygone era, this 'sarcophagus' has become the antique of the future. After the Great Wall of China and the wall of the Atlantic, man has just laid the foundations for a Wall of Babel, and this wall is not to become a tower anymore, like the one once thrown up to enable man to reach to heaven. This one's purpose will be to try and prevent 'the fire from heaven' from coming down to ravage the Earth.[4]

If war once belonged to 'the art of laying siege', peace is now hoping to break out in the art of laying a bunker, a rampart against 'the state of siege of Progress'.

Yet, on 16 September 2003, the US Senate, with its Republican majority, rejected an amendment introduced by the Democrats that aimed at stymieing research and development of nuclear weapons 'suited to destroying bunkers'.

This programme of unconventional weaponry, known as the 'bunker-busting bomb', is supposed to allow penetration of underground shelters held by regular armies – and also by terrorist groups.

According to the calculations peculiar to the American military planners, there could be some 10,000 shelters of this kind distributed throughout the world.

Let's not forget that the House of Representatives had previously adopted a bill allocating nearly eleven million dollars for the building of the nuclear weapons plant and an additional five million for work on designing the 'penetration vehicle,' the Robust Nuclear Earth Penetrator, (RNEP).[5]

Here is the ultimate or, more precisely, the penultimate duel between 'the sword and the cuirass'. As for the absolute ultimate, this is it: David Stevenson, an American researcher at the California Institute of Technology, Pasadena, in the journal *Nature* of 15 May 2003, proposed speeding up the China syndrome – in other words, to reproduce the threat of the 1979 Harrisburg accident, and to do so in order to 'to deepen[!] our knowledge about the Earth's core.'[6]

To reach the 'Earth's centre', Stevenson proposes nothing less than enlarging a very deep fault line, 6,000 km long, by producing a series of underground nuclear explosions, corresponding to the equivalent of an earthquake that would measure seven on the Richter scale.

This time it would be a matter of penetrating the Earth's crust as far as the inner core, whereas till now geological forages have never gone beyond a few tens of kilometres and our knowledge of the lithospheric mantle scarcely goes beyond 300 kilometres. As for the 'penetration vector' that would serve as a probe, this would be composed of a 10,000 cubic metre chunk of cast iron, which translates to 100,000 tons in weight. This ball or, rather, this 'cannon ball' would disappear, they tell us, at a speed of 18 km per hour, into the bowels of the Earth . . .[7]

Throughout the history of conflicts, the lance and the sword have come up against the shield, and the rain of arrows against knights' armour, just as the cannon ball and then the explosive shell have demolished the enclosing walls of citadels. As for the bomb, it has had a hand in burying troops in deeper and deeper casemates. This has gone on right up to the invention of the atomic weapon, whose penetration capabilities have only been limited till now by deterrence and, in particular, by the banning of atomic tests.

But since 2001 everything has changed, for the exterminating radiation of the neutron bomb has just been outclassed by a penetration capability that no longer penetrates the air space of geostrategy 'full-scale', but 'full-depth', within a lithospheric mantle that then becomes the ultimate megalithic wall, the last sarcophagus of humanity.

And so, against the Chernobyl time barrier where the architectonic resistance of concrete protects us from radionuclides left over from the year 1986, they are getting ready to shortly pit the antitelluric power of a cannon ball capable of perforating not only the resistance of an old building material, reinforced concrete, but the actual tectonic resistance of the geological plates that constitute the Earth's structure.

Against the tellurism of the 'great volcano' and its prehistoric ravages, the twenty-first century man of science is gearing up to pit the antitellurism of the militaro-geological atom, thereby turning nuclear energy into the all-purpose energy of a fanatical demiurgery, the colossal ecological havoc of which was, after all, demonstrated by the Soviet catastrophe.

Radioactivity of the contamination of the future or radiotoxicity of a science without a conscience that is no longer merely the 'ruin of the soul,' but the ruin of the space-time of a unique material: that of this habitable telluric planet; this 'fullness' that still protects us from the cosmic void that some boast of conquering, while others, just as determined, are getting ready to pierce the mysteries, even unto the centre of the Earth, without giving a moment's thought to the risks run.

'The weapons scientists have become the alchemists of our times, working in secret ways that cannot be divulged, casting spells which embrace us all,' Solly Zuckerman reckons. 'They may never have been in battle, may never have experienced the devastation of war, but they know how to devise the means of destruction.'[8]

From the arsenal of Venice in the age of Galileo right up to the secret laboratories of the post-Cold War, via the Manhattan Project of Los Alamos, science has become the arsenal of major accidents, the great catastrophe factory toiling away in anticipation of the cataclysms of hyperterrorism.

'He that deviseth to do evil shall be called a mischievous person,' spelled out the Book of Proverbs (Proverbs 24:8). What can you say about the supreme mischief that consists in hijacking not only planes and vehicles of all sorts, but in hijacking the 'great vehicle', the whole set of knowledge, in physics as in biology or chemistry, to achieve, ultimately, the greatest possible amount of terror?

Once more according to Martin Rees, since the middle of last century or, more precisely still, since the business of the Cuban Missile Crisis in 1962, the risk of a worldwide atomic disaster has risen to 50/50. But now, this familiar risk so often wheeled out to justify the endless relaunching of the arms race, is topped off with the growing threat, as we have just seen, of untimely discoveries of a magnitude that exceeds all rationality.

Listen now to the colourful tale of the bomb-disposal experts of the Soviet army, covering the period 1945–50.

> Our unit was not going to be dissolved: we were going to clear the fields of mines; the land had to be handed back to the peasants. For everyone else the war was over, but for us bomb-disposal experts it went on. The grass was high, everything had shot up during the war, it was hard to hack out a path, when there were mines and bombs everywhere all around us. But the people needed the land and we went as fast as we could. Every day, comrades died. Every day, someone had to be buried.[9]

This is the account of a woman who was a soldier in the Corps of Engineers, talking to Svetlana Alexievitch, herself a witness to the nuclear accident of 1986 . . . But, at Chernobyl, it was no longer the bomb-disposal experts who were sacrificed, it was the Earth! To bury the ground – even the theatre of the absurd would not have dared such an apocalyptic pleonasm.

Currently, what is undermined and everywhere contaminated is science, the whole set of our knowledge literally poisoned by an arms race in 'weapons of mass destruction' that is infesting what we learn and will, if we are not careful, shortly decommission science, making it unavailable to do good.

Tomorrow, hot on the heels of Mother Earth, maybe 'science', this 'wisdom' deriving from a knowledge that was, though, the distinctive feature of *homo sapiens*, will also have to be buried.

'You make war with weapons, not with poison,' decreed the Roman law-makers. Driving this observation home, the then United Nations Secretary-General, Sithu U Thant of Burma, declared in June 1969: 'The notion of hostilities being out of control is incompatible with the notion of military security.'[10]

We know the rest, with the exponential development of biological and chemical 'weapons' that threaten humanity every bit as much as nuclear weapons do.

Relying on purely military arguments and not on moral considerations, the Secretary-General of the United Nations concluded with these words: 'The very existence of these weapons contributes to international tension without offering any obvious military advantages in compensation.'[11]

Some thirty years on, the prophecy has been fulfilled by this terrorist hypertension that totally perverts international politics.

Indeed, if mass destruction is within reach of the socially excluded, the argument for deterrence evaporates and we are at the mercy of any and all catastrophes, catastrophes either deliberately triggered by clandestine groups or industrial or other 'major accidents'.

One forgotten example of this, among others, is the discovery made at Denver Airport, by a Democrat deputy, of a depot of 21,108 missiles, each formed by a cluster of 76 gas bombs, the whole lot at the mercy of a fire. The capability of this hidden arsenal: extermination of the entire population of the globe.[12]

But let's get back to Russia and the splendid offer made to young victims of the Chernobyl nuclear catastrophe: 'Since 1995, a decree of the Ukrainian government has ensured that children from the contaminated zone have been offered seaside holidays all along the Crimean coast. Using magneto-therapy, aromatherapy, and so on, the cure proposed has the special merit of allowing these children of the final shore to discover the joys of the beach.'[13]

To make their summer stay as cheery as possible, the old naval and former top-secret military base of Kazachya Bay, a training centre for mine-carrying and -monitoring dolphins, has been reconverted into an aquatic circus, a Marine Land of the Big Night.

'God has acted wisely by putting birth before death; otherwise, what would we know of life?' wrote humorist, Alphonse Allais.

Since then, this humour that that other humorist, Pierre Dac, somewhat abused, has been turned on its head. For now the origin of life or, more precisely, of humanity's survival, is the all-out search for death. Not the death of the other, of the enemy or of some kind of adversary any more, but the death of all in the suicidal state of mutually assured destruction. If that is not the theatre of non-sense, it looks horribly like it. So much so that we would need to turn Alphonse Allais' aphorism around and write: 'The demon of nonsense has acted wisely by putting the end of life before the beginning; otherwise, what would we know of non-life?'

Indeed, as we have just seen, the knowledge at issue today, in the laboratories of 'advanced' research, no longer involves mere externalization, outsourcing, the fine-tuning of an eccentric unemployment, but extermination, the end of everything – in other words, life in reverse.

And so, little by little, foreknowledge of the end has invaded scientific thought, before extending to the political economy of a globalized world.

It is here and now that one of the most controversial questions in the history of knowledge is posed: the question of a possible doping of technological culture, meaning scientific thinking as a whole.

'Doping': no longer targeting the muscular prowess of the athlete's body as he or she is dragged kicking and screaming into the craziness of some boundless perfectibility, but targeting military knowledge regarding power, the death instinct; that militarization of science that has recently wound up in the ruin, not of the soul, but of a scientific spirit dragged along by the absurd perspective of the supremacy of the death principle. This began with the explosion of the atom bomb and has continued right up to the designing of the future genetic bomb via this information bomb that will have fuelled the comprehensive blasting of common sense.

With these 'zero sum games', Olympic games of a kind produced by a fatalism that taints technoscientific thought, we can more easily understand the accident in knowledge that today rounds off the accident in substances, in a world that is now a victim of terror, with the tacit consent of far too many savants.

As Jean-Pierre Vernant explained in August 2004: 'Modern sport is bound up with the idea of indefinite progress in the technics of the body, in the tools that the different trials can use and in the human being's capacity for excelling him- or herself and for always improving on their scores.'

Vernant concludes, apropos the Athens Olympic Games: 'The notion of a record has no place in the olympiad by its very nature. What matters is winning, not doing better than your predecessors, not only because we still don't have the technical means of measuring time exactly, but because the idea that sport constitutes a form of activity that can be perfected indefinitely does not and cannot exist.'[14]

Actually, the 'progressivist' belief in the possibility of social progress popularized by the Great Movement of the nineteenth century is behind all the competitions, whether political, economic or cultural, of the modern industrial age – right up to the unbridled competition that is the basis of the contemporary 'turbocapitalism' of globalization.

Whence the scale of the phenomenon of doping and performance-enhancement of the global economic system, way beyond the sports stadia and right into the more or less clandestine dispensaries of the transgenic biotechnologies. As one writer points out, drawing his inspiration very broadly from the concept of 'mimetic desire' forged by René Girard: 'With media coverage of competitions, all-out performance-enhancement is uncontrollable. If opponents were till now simple obstacles to the achievement of the "desire for victory", it is as the obstacles they represent that they are now valued. The "desire for an obstacle" has taken over and so what is now sought is adversity, not the adversary.'[15]

The writer goes on to say of this mimetic condensing:

> In such conditions, the sports show will move towards being set to images, a process based on an entirely new dramatic art that will allow opponents to get 'burned'. [. . .] So it is likely that, as far as doping goes, the worst is yet to come. If nothing is done about it, a sort of pathological desire will soon structure the whole process of access to victory, sweeping opponents along towards their own destruction.

After the 'Olympic games' of Antiquity, the survival games of the human race in the age of nuclear deterrence have outrageously amplified this mimetic pathology. But here, what is about to get 'burned' or, more precisely, vitrified, is no longer the adversary who vanished into the East. It is adversity, with these faceless and homeless terrorists hell-bent on collective suicide.

They say that in the United States law professors have been arguing, for some little time now, that 'if torture is the only way to avoid the explosion of an atomic bomb in Times Square, it is licit.'[16]

After all-out deterrence, the extension of the torture chamber is thus once more on the agenda for the day, the last day . . . For if everything is allowed in order to avert the end of the world, then it is the end of everything!

The end of law, including the law of the fittest, trumped by the law of the maddest: what is required is to urgently reopen the camps, all the camps, not only the camp at Guantanamo Bay, but those of Treblinka, Auschwitz and Birkenau, in order to finally get ready for what André Chouraqui used to call, not so long ago: the 'planetary shoah'.[17]

10 The Dromosphere

A quarter of a century ago, in 1978, Federal Germany road-tested a revelatory experiment: removing all speed limits on the autobahn. Organized conjointly by the government, the car manufacturers and the car clubs, this series of tests and sundry investigations was designed to get past old hat analyses of the causes of car accidents. Everyone suddenly put forward factors that had been overlooked: the state of road surfaces, atmospheric conditions, and so on and so forth. These private and public bodies suddenly seemed to join forces to deny that speeding was directly responsible.

According to this lot, speed was neither the sole nor even the main cause of road accidents and their seriousness; other factors carried greater responsibility in the carnage caused by automobile transport.

As we might suspect, the real reason for such an about-face lay elsewhere. According to the German car makers, 'To condemn vehicles designed to travel at 150 km or 200 km an hour to do only 130 is to condemn technical progress and thereby the position of German industry on the foreign markets, thus opening the floodgates to unemployment.'

In the face of this speech for the defence, the federal government decided to 'free up the autobahn'. Even though drivers were recommended not to exceed 130 km per hour, doing 200 km or 250 km per hour was not to be penalized any more, car drivers' self-discipline was to suffice . . .

Anxious French car makers of the day were to come up with a complementary argument: 'On the highway as in competition, the more a car is built to go fast, the more reliable it is. When you can do more, you can do less.'

As it happens, competition on the foreign markets, notably in the United States where German cars sell extremely well, is not about speed, which is strictly limited. What counts is reliability, which is a function of the maximum speed even if this is rarely used. Germany having deliberately opted for 'a few more dead today, fewer unemployed tomorrow' competitiveness, declared the French car makers, is driving us down the same road.

We know what ensued as far as mass unemployment goes, in Germany as in France and elsewhere. What is revelatory about this period, though, is the acceptance of the notion that road victims are victims of progress. From that moment, every car driver becomes a sort of 'test pilot' of cutting-edge technologies. Those who risk their own lives and the lives of their fellows can know that they are putting them at risk in order to ensure the reliability of the product, the smooth running of the national firm – in other words, security of employment.

Since the advance of the car industry appears assured and guaranteed by excessive speed, to risk your life for the security of speed is equivalent to risking your life for the employment of time and no longer, as in days gone by, for the homeland, in defence of the employment of the national arena.

With this form of time management, which is curious, to say the least, and which is shored up at once by social security and civil security which list the work-place accident and the accident to and from work under the same heading, it is no longer a matter, as it was in the past, of covering up an accident or failure, but indeed of making it productive, psychologically speaking.

This process aimed at triggering a sort of deregulation of behaviours already heralded the coming age of full-scale deregulation in which we now live, following the self-regulation of traditional societies and the regulation of institutional societies.

Like the Russian people once called on to make sacrifices to ensure the 'radiant future' of a scientific communism, the technical progress of capitalist societies was to be indexed to the sacrifice of consumers.

Strangely, in those not so distant days of the 1970s, when technoscientific progress was assimilated to the risk of driving a fast vehicle, the French government of Raymond Barre insisted, only the day after the Three Mile Island catastrophe of 1979, on the need to speed up construction of French nuclear power stations, thereby moving in the direction of this eschatological perspective.

From that moment, were we really to seriously envisage the rise of an officially cynical, meaning purely sadistic, power? The advent of a 'suicidal national state', not so much political as transpolitical, which would soon be exemplified by the Chernobyl catastrophe with the implosion of the Soviet Union?

It might now be useful to dust off an old short story of Ursula Le Guin's: 'Direction of the Road' (which appeared in France precisely in 1978 under the title of 'Le Chêne et la Mort' – the oak and death).[1]

In this fictional tale, the author gives voice to a tree more than two hundred years old that grew up with the accelerated gallop of horses pulling diligences and, shortly after that, the acceleration of cars, right up to the fatal accident that brings the tree's first-person narrative to a close.

In the days of horses, claims the oak, 'they [the horses] did not used to be so demanding. They never hurried us into anything more than a gallop, and that was rare.' But then the first motor car appeared, and then another: 'a new one, suddenly dragging me and the road and our hill, the orchard, the fields, the farmhouse roof all jigging and jouncing and racketing along from East to West. I went faster than a gallop, faster than I had ever gone before. I had scarcely time to loom, before I had to shrink right down again.'

Building on this dromoscopic vision, our oak goes on:

> But have you ever considered the feat accomplished, the skill involved, when a tree enlarges, simultaneously yet at slightly different rates and in slightly different manners, for each one of forty motorcar drivers facing two opposite directions, while at the same time diminishing for forty more who have got their backs to it, meanwhile remembering to loom over each single one at the right moment: and to do this minute after minute, hour after hour, from daybreak till nightfall or long after?
>
> For my road had become a busy one; it worked all day long under almost continual traffic. It worked, and I worked. I did not jounce and bounce so much any more, but I had to run faster and faster: to grow enormously, to loom in a split second, to shrink to nothing, all in a hurry, without time to enjoy the action, and without rest: over and over and over.

Our venerable oak sets himself up as 'an oak of the law':

> For fifty or sixty years, then, I have upheld the Order of Things, and have done my share of supporting the human creatures' illusion that they are 'going somewhere'. And I am not unwilling to do so. But a truly terrible thing has occurred, which I wish to protest.
>
> I do not mind going in two directions at once; I do not mind growing and shrinking simultaneously; I do not mind moving, even at the disagreeable rate of sixty or seventy miles an hour. I am ready to go on doing all these things until I am felled or bulldozed. They're my job. But I do object, passionately, to being made eternal.

There then follows a detailed description of an accident in which a driver crashes into the oak: 'I killed him instantly [. . .] I had to kill him,' the tree freely admits.

> What I protest, what I cannot endure, is this: as I leapt at him, he saw me. He looked up at last. He saw me as I have never been seen before, not even by a child, not even in the days when people looked at things. He saw me whole, and saw nothing else – then, or ever.
>
> He saw me under the aspect of eternity. He confused me with eternity. And because he died in that moment of false vision, because it can never change, I am caught in it, eternally.

As the philosophical oak explains, by way of conclusion:

> This is unendurable. I cannot uphold such an illusion. If the human creatures will not understand Relativity, very well; but they must understand Relatedness.
>
> If it is necessary to the Order of Things, I will kill drivers of cars, though killing is not a duty usually required of oaks. But it is unjust to require me to play the part, not of the killer only, but of death. For I am not death. I am life: I am mortal.

Exhuming this text that is more than quarter of a century old may seem anachronistic today, but that would be a mistake or, rather, an optical illusion produced by the acceleration of the real.

Indeed, at the beginning of last century, three million plane trees, maples and poplars still lined the roads of France, but now there are only 400,000 of them left, and these are held responsible for 750 deaths a year.[2] The notion of fate that still prevailed half a century ago was superseded by the principle of collective responsibility. And so the concept of the 'unforgiving road' was born.

At the Department of Roads, the statisticians calculated that there was four times the risk of dying in an accident running into a tree than in any other type of accident.

The way they put it, the magnificent leafy monarchs of the plant world have become a potential minefield in terms of human lives, whence their sacrifice, tabled in a 1970 circular urging their systematic eradication.

In 2001, the Minister for Agriculture, Jean Glavany, was still hoeing into plane trees as public menaces.

Of course, there are those who dare to assert that 'it is not plane trees that cross in front of cars,' but what can you say to those who go as far as rescinding the abolition of the death penalty in order to justify the felling of lateral obstacles, considered aggravating factors in any road accident? 'Certainly,' they say, 'every driver should remain in control of his vehicle. But, all the same, death is too heavy a price to pay.'

The die has been cast. Between yesterday's fiction and today's accelerated reality, the difference has disappeared and, with it, all reason. As an admission of helplessness, one regional councillor even came up with this: 'What do you want us to do? We'll never get them to slow down!'

Speed suppresses not only Relatedness, as Ursual Le Guin so acutely explained, but also Reason. This is what must finally reveal the importance of the accident in contemporary thought, in other words: the accident in the circulation of knowledge between 'being' and 'place', this backdrop to life that comprises not only the animal realm – that of the being's movements – but the plant and mineral realms, that is, the realms of stability, fixity and, finally, the persistence of sites.

How long before we see the elimination of hills and cliffs, the definitive levelling of the world's relief?

How long before we see the abolition of the waves of the high seas, of this set of collateral obstacles that still put the brakes on the acceleration of technical progress?

When you run into a table, should you do away with it or learn to avoid it?

Since we have, seemingly, erased distances, it remains to eliminate the resistance of materials, of lithospheric or hydrospheric elements.

Just as we were able in the recent past to go beyond the whole set of atmospheric elements, thanks to the velocity of escape from gravity, from terrestrial weightiness, we must now eliminate what still subsists of material opposition to advancement, to the dromospheric race of automotive devices.

After the silk weavers of Lyon and the machine-wrecking Luddites of Britain, the time of those motorists bent on systematically felling the greenery, bringing down the shade with it, replaced as it now is by motor vehicle air conditioning, is now upon us . . . How long before the four seasons are eliminated and replaced by the single temperate climate of a general planet-spanning air conditioning system? How long before the meteorological atmosphere is put under glass or, rather, under sequestration, thanks to this sphere or, more precisely, this dromosphere, in the race of a progress that is nothing more than a third type of inflation, not so much economic as eschatological, since with it the acceleration of reality once and for all shunts all historical accumulation aside?

'By dint of wanting to possess, we are ourselves possessed,' noted Victor Hugo two centuries ago.

No point looking any further for the origin of this hyperviolence that is now unfurling all over the world, for speed has become the very quintessence of such violence, eliminating one by one any markers, not only any 'temporal' markers but also any factual limit.

Take the example of statics and of the resistance of materials that are the basis of any construction: since last century, for instance, we have been going faster and faster to achieve the durable, the durability of those buildings and structures of very long duration that condition the permanence and stability of our societies.

As an architect put it after the collapse of the Roissy airport terminal in May 2004: 'Building sites have to go faster and faster, technical performances have to be more and more precise, to the point of verging on extreme complexity. We could even talk of an "ideology of speed" and of performance.'

No, dear colleague, it is not even a matter of some passing ideology any more, as it was at the beginning of the industrial age, but of dromology. And this is worse, since it conditions the whole of technically-oriented civilization, just as the historian Marc Bloch told us it would.

Speaking of which, it would perhaps be appropriate to shift the concept of surrealism as heir to the domains of art and literature, and apply it to the field of politics. Sylvie Guillem, the dancer, does this, for example, when she declares: 'You have to dance, not over-dance.' In other words, content yourself exclusively with choreographic feats.

Actually, as soon as these feats, these technical, scientific or industrial achievements, are wholly conditioned by the acceleration of the real, one can just as easily over-build as over-destroy.

Whence the all-out spread of a surveillance – over-vigilance – that surpasses the 'state of vigilance' of those not so distant days 'when people looked at things,' as Ursula Le Guin helpfully reminded us. And this merely anticipates the imminent, or practically imminent overdone or *super*-humanity that the apostles of Progress are cooking up for us in the secrecy of the laboratories of transgenic genesis.

'People bewail effects, but make the most of causes,' wrote Bossuet. Paraphrasing another giant of critical writing, we might echo: 'If science wants nothing to do with its effects, ignorance will get hold of it.'[3]

These days, curiously, the sphere of acceleration of reality tends to reverse the principle of responsibility.

With the tree that kills, the reality of guilt is transferred from the guilty to the innocent, the innocence of a vegetal fixity that creates an obstacle foiling the automobility of a vehicle which is more often than not, now, driven with the assistance of a computer.

Here, the dromoscopy – this optical illusion of flashing past that reverses the direction of the road, with its trees that look like they are hurling themselves at the windscreen before vanishing in the rear mirror whereas, in reality, it is the reverse that is happening – affects the whole gamut of our perceptions and muddles our judgement to the point where the victim suddenly becomes the designated guilty party.

Strangely, too, this phenomenon of dromoscopic transfer today affects our legal system without this worrying the authorities too much. I'm talking here about the transfer of guilt that is overturning a number of trials in the criminal courts where the victim of the crime statistic subtly morphs into the guilty party. This permutation is, doubtless, an indirect consequence of the too-great mobility of viewpoints in the endless acceleration of our social behaviours.

Look at what is happening, for example, in the corporate firm subject to economic globalization: as soon as employers have a serious problem, they transfer, relocate, and the more delicate the commercial situation appears, the faster they tend to act. This even involves the phenomenal expatriation that is turning corporate life on its head.

If perception eclipses the reality of the moment to that extent, this is because there is no more intermediary financing, no more extensions of time, no more intervals of intervention.

Reducing to nothing the space-time of our actions and our interactions, acceleration suddenly upends the reality of the facts. And so, the dromosphere induces everywhere at once an illusory reversal of our know-how and our acquired knowledge, whereby temporal compression of our activities illustrates very precisely what Aristotle called 'the accident of accidents'.

For instance, what happens during acceleration of automobility, that is, where what stays put appears to flee while the interior of the vehicle seems immobile, is reproduced today in the media perception of televized reality.

What I baptized DROMOSCOPY twenty years ago[4] now applies to the whole of our experience and knowledge, the fruits of a now-remote era, one so slow that these gains seem to flee in turn, discredited by the mad race of contemporary events.

Whence the inertia of the present tense, known as *presentism*, which is nothing more than the illusion of acceleration in communications, the telescoping of a fleeting teleobjectivity that tends to replace established objectivity, the same way the dromoscopic illusion of the automobile disturbs and seriously perturbs the roadside, making an immobile environment appear mobile by providing those in the vehicle with the comfort of an on-the-spot fixity that only an accident along the way will jog them out of, a head-on collision suddenly re-establishing the facts.

In this instance, the fixity of the obstacle rears up like some Justice of the Peace of the shift in perspective, and the tree or wall are only ever features of what the obstacle of the geophysical finiteness of a unique habitat represents, further down the track, for a species of animal every bit as much 'rural' as 'human'. It is a habitat that no transgenic engineering will make us leave, despite the postmodern ranting about some virtual space, a surrogate sixth continent for a neocolonialism every bit as illusory, in the end, as the conquest of astrophysical space by the adepts of NASA's 'manned flights' in the 1960s.

Indeed, what astronautical illusion did yesterday's moon missions embody? What conquests, what 'fallout' was it a matter of, then, if not that of a space indefinitely travelable but uninhabitable!

In other words, the fallout of a cosmic vacuum bearing no relationship to biospherical space, where what is travelable is simultaneously inhabitable, where circulating and settling are one and the same 'abode'.

To so unduly privilege exotic feats to the detriment of any 'dwelling' – now there's a crazy act for you, an action of panic deterritorialization that only 'the balance of terror' between East and West could provoke in the face of the probabilities of an atomic war making the Earth definitively unfit for life.

And so, the so-called 'conquest of space' was merely confirmation of Bossuet's observation, the cause of such 'exotic' progress only ever having been the effect of terrorist deterrence between communism and capitalism. As a naval officer recently put it: 'Surely a successful military manoeuvre is a catastrophe averted just in time.'

Nothing is ever gained without something being lost, and, therefore, technical progress is only an agreed sacrifice; proof of this was offered to us yet again by the launching of deep-space astronautics during the Cold War years.

At this period in history that at one point saw the threat of the Soviet nuclear missile installation on Cuba (1962) put the all-too-precarious balance between the two great blocs at risk, planet Earth confronts a major hazard whereby, according to the astrophysicist, Sir Martin Rees, 'the odds are no better than fifty–fifty that our present civilization on Earth will survive to the end of the present century . . .'[5]

This was confirmed by the historian Arthur Schlesinger, former special assistant to President John Kennedy, who claimed in his memoirs on the subject of the Cuban Missile Crisis that: 'This was not only the most dangerous moment of the Cold War. It was the most dangerous moment in human history.'[6]

So this is it, the successful manoeuvre: the conquest of space resulting from the catastrophe of the sacrifice of the planet averted just in time, in the duel between East and West! Strangely, such a 'military' manoeuvre seems to be enjoying a comeback, with the Pentagon installing the first missiles of a future 'anti-missile belt' during the summer of 2004, in a rush to see them in place for the presidential elections of 2 November 2004. And this was done without any conclusive experiment verifying the effectiveness of the system. Similarly, in his election programme the incumbent president, George W. Bush, did not budge from the course set at the beginning of the year for astronautics. And so, when the International Space Station (ISS) is complete, a new craft is set to effect its first manned mission in 2014, before taking Americans back to the moon some time between 2015 and 2020. As for Bush's Democrat opponent, John Kerry openly slammed these pointlessly costly objectives and offered no set goal and no set calendar for space exploration.[7]

Heralding a reversal in perspective, Nietzsche wrote: 'Love your furthest away as you love yourself.'

In the United States, this azimuthal projection seems to be back in pride of place with the saying: 'To annihilate the enemy close to you, you have to first strike the one further away.'[8] Whether a preemptive strike at the end of the Cold War or a preventive war on terrorism today, the same 'forward-scatter' logic has long been at work.

Dromological logic of a race for 'all-out' supremacy that causes our nearest and dearest to disappear in favour of the furthest away, all that is furthest away, all the exoticisms, in other words, every manner of exodus!

A race beyond Good and Evil that renounces all the 'on-this-sides' only to wind up, ultimately, at this topological reversal whereby the *global* now represents the interiority of a finite world, and the *local*, its exteriority, that great suburban belt of a history without geography – a *chronosphere* of present time, 'real time', that has replaced the *geosphere* of life's arena.

This, admittedly, is the conclusion of Bossuet's sage little phrase: 'People bewail effects, but make the most of causes.' There is one qualification, though: it is the 'weak' who bewail the disasters of technological progress and the 'powerful' who most readily make the most of causes.

The excluded are exiled on all sides; for them the globalized foreclosure ends in equally all-out exclusion. No need here to wheel out once again the great transcontinental migratory flows of dire poverty.

What can you say, for example, about these old retirees that never stop travelling around the world trying to see it all before signing off, while no one at all bewails the idle young who have already seen it all before beginning to live?

In the end, the progressive pressure of the dromosphere is nothing but a headlong rush that leads to this externalization – outsourcing – that is only ever the postmodern term for extermination. Revelation of a finiteness in which, globality disqualifying all locality, Hegel's *schöne Totalität* appears for what it is.

After two millennia of experiments and failures, of accidents of all kinds, with globalization the third millennium inaugurates the paradox of the failure of success for it is the success of Progress that provokes disaster. An integral accident of a science now deprived of a conscience, whose arrogant triumph wipes out even the memory of its former benefits.

This is a major event in a long history of knowledge whose tragic nature globalization both reveals and conceals at one and the same time.

After that, it is not so much error, some system failure, or even large-scale catastrophe that brings the boom in knowledge to a close, but the very excessiveness of the feats achieved in the face of the limits of a cramped planet. As though, in the course of the past century, the promotion of technoscientific progress had doped science, as certain banned substances do the body of the athlete. So here it is no longer the congenital weakness of the various branches of knowledge that is the limit, but indeed the power of a science that has become a 'hyperpower', in this race to the death represented not so long ago, with the arms race, by the militarization of science.

A fixed limit, this one, for it is the product of a galloping success that no one really contests, but that brings down knowledge still based, only recently, on the humbleness of an experimental know-how, a minor branch of learning at the origins of scientific reasoning, that has now become major due to the inordinate scope of its impact, of its panic-inducing results.

Here again, the example of doping in sport is useful, it would seem, to the demonstration: what value, in fact, can progress have when it not only denatures but literally exterminates the person or people who are, they say, its 'beneficiaries'?

Disaster of a contagious progress that the limits of the world can no longer bear, any more than the set of living beings can. In fact, and contrary to the failures of experimental know-how, the disasters of progress can no longer be overcome as the failings of a totally new knowledge were, in the past, in that age that is not as distant as all that, where the modesty of genius still allied 'science' and 'philosophy'.[9]

But let's go back over the phenomenon of the acceleration of reality, so perceptible today in the retooling of nations' foreign policy.

In recent years, the United States has decided to see international conflicts as internecine wars, between states considered more or less rogue. But in this extroverted world, rapid deployment of US armed forces has hoodwinked and abused the United States in the most peculiar way about the reality of its hegemony.

Just as adaptation of the eye is a function of the car driver's speed, the optical point expanding in the distance with acceleration of the vehicle, today the geostrategic perception of America as a hyperpower has moved beyond the limits of the United Nations to embrace the curve of the globe.

For the Americans the dromosphere is thus no longer a metaphor of progress but an avowed fact of their geopolitical perception in which topological reversals become more and more frequent.

Recently, a wit asked himself this question: 'How can American society, so wealthy and so multicultural, project itself in such a monolithic way? There is a lot of talk of an American empire, but it's an empire behind a barricade. The Americans are in bunkers.'[10]

Yet the answer is simple, even simplistic. This monolithic state is no longer that of a 'totalitarian' power comparable to those of the recent past; it belongs to the definitive closure, the foreclosure, of the world.

This 'globalitarian' perception is thus indeed that produced by the dromosphere tightly embracing the ultimate curve of the terrestrial star. Here dromoscopy reaches its apogee since its horizon is no longer the line that once separated sky and earth, but only the geodesic curve that distinguishes the full from the empty. 'Biospherical' fullness, of this 'exospherical' finiteness, this intersidereal milieu that even conditions terrestrial volume, since 'any limit comes from outside' and the spherical form of celestial objects comes from their perpetual motion – in other words, how fast or slowly they rotate.

In the face of this 'big bang' that no one seems to be turning a hair over, the famous monolithic bunker is never more than a cloister: the cloister of history.

Having attained the perfection of its orbital circulation, the dromosphere thus brings to a close the age of political revolutions in order to peek into the Pandora's box of transpolitical revelations. This is perhaps the essential part of André Malraux's intuition about the twenty-first century.

By way of confirmation of this 'historic tellurism', let's hear it from French columnist Thomas Ferenczi, in an editorial devoted to the expansion of the European Community:

> Europe is also internal politics. Once European politics and national politics start to overlap more and more, can we feasibly separate them when it comes to voting, without going so far as to pit a Europe of the Left against a Europe of the

> Right, at the risk of caricaturing Europe altogether? Don't we have a right to expect a certain continuity? . . . What is at stake in such controversies is French politics in its European dimension.'[11]

What our editorialist doesn't see is 'the crisis in the concept of dimension.'[12] The crisis in these whole dimensions, at once geometric and geopolitical, that today leads to the fractalization of the concept of identity (national, communal) and so to this 'critical space' where nothing is whole any more, apart from this 'astropolitical' sphere that no one dares conceive of, except perhaps the Little Prince! This is, in the end, the age of revelations that succeeds the age of clapped-out rotating revolutions that the past century literally exterminated through the extravagance of its accelerated 'progressivism'.

In a prophetic interview, Alain Rousset, the president of the regions of France (ARF), declared: 'Society is anxious. It no longer seems able to project itself into the future, to imagine that tomorrow might be better than today. We need to reflect on how to get back to the idea of progress. This idea is one of those most closely tied up with the Left, for the same reasons as justice.'[13]

What Rousset is referring to here, as we might suspect, is obviously not only the parliamentary left, for the obstacle of finiteness far outstrips the democracy of political assemblies. In the jargon of the Department of Civil Engineering, for instance, trees and anti-noise barriers and even security ramps along the highways are currently known as 'lateral obstacles'. What can you say of the – this time head-on – obstacle constituted by the geodesic curvature for those who still claim to be 'going with the flow of history'?

These are upholders of a historical materialism that turns its nose up at any geophysical materialism, even if very evident. Internationalism of the proletariat yesterday, turbo-capitalism of the single market today, the day comes when the star can no longer bear the disaster of progress, the collateral damage that results, as we have seen throughout this book, from the acceleration not only of the history of humanity, but of all reality.

In fact and for the first time perhaps in such a tangible way, for each and every one of us the perimeter of life is strictly circumscribed by the void. The old fullness of the biosphere has been overtaken, now, by this negative horizon that defines both the world and what is out of this world at once.

'Outside is always inside,' crowed the architects of yesterday's triumphant modernity. From now on, outside is exodus, the exosphere of a space unfit for life.

By way of confirmation of this admission of failure, note the astronomical search for exoplanets way beyond the solar system. For telluric planets, as they say, to describe terrestrial-type stars that are at once small and solid.

Not an easy quest, since no extrasolar planet likely to harbour life has yet been spotted, all those so far catalogued only ever being gigantic bubbles of gas too scalding hot to favour the chemistry necessary for a zone of habitability conducive to the emergence of life.[14]

Despite this, the *Astrophysical Journal* announced, at the end of the month of August 2004, that American researchers had just come across one! But we were later to learn that the three exotic planets in question had a mass fourteen to twenty times greater than that of our dear old Earth. Yet again, in this 'super-earth' race, the United States were hoping to revive the old myth of the 'frontier', a Far West that has nothing to do any more with the 'forward-scatter' model of the pioneers of transhumance at all costs. It is now all about a transhumanity exiling itself in quest of a vaster earth, the promised land of a new 'New World', one no longer lying westwards across a continent, but over our heads, in the firmament!

After the collapse of the New York skyline in 2001, something else had to be found urgently to stretch out ever further the American dream, the myth of a destiny manifest in the United States. The preacher, Billy Graham, expressed this clearly in his sermon of 14 September 2001: 'But now we have a choice: whether to implode and disintegrate emotionally and spiritually as a people and a nation – or whether we choose to become stronger through all of this struggle – to rebuild on a solid foundation.'[15]

After the Soviet Union, is the United States in turn going to implode and disintegrate before our very eyes like the Twin Towers?

Or are we going to see the exotic refounding not only of America, but of the United Nations?

At the end of the road, is humanity going to wind up finally taking off, becoming an Unidentified Flying Object, as New Age pundits or the survivalist sects springing up like mushrooms all over the United States would have us believe?

If globalization is certainly not the end of the world, it is nonetheless associated with a sort of 'voyage to the centre of the Earth', to the centre of real time that has so dangerously replaced the centre of the world, that space, undeniably real, that always used to organize the intervals and time limits for action – before the age of widespread interaction.

Everything, right now! Such is the crazy catch-cry of hypermodern times, of this hypercentre of temporal compression where everything crashes together, telescoping endlessly under the fearful pressure of telecommunications, into this 'teleobjective' proximity that has nothing concrete about it except its infectious hysteria.

Let's not forget: too much light and you get blindness; too much justice and you get injustice; too much speed, the speed of light, and you get inertia, polar inertia.

Following on from the ancient observation of the impact of atmospheric pressure on meteorology, surely it would be appropriate to pinpoint, finally, the havoc now wreaked by dromospheric pressure, not only on history and its geography, but on the political economy of a democracy now subject to the DROMOCRACY of machines, machines for producing systematic destruction, that are now indistinguishable from war machines.[16]

By way of illustration of this insanity, we might cite one last anecdote. In the United States they are apparently bottling the planet: derived from some old NASA research, the ecosphere is a simplified version of our ecosystem. It is also the latest gadget, the very latest folly in the realm of interior decoration. Locked in a glass bubble like a fishbowl, this model of the atmosphere has a lifespan of two years. An optical illusion, whoever acquires it becomes master of a scale model of the world.

Notes

Chapter 1 Caution

1 Paul Valéry, La Crise de l'intelligence in *Cahiers* (1894–1914) (Paris: Gallimard, 1987), vol. I.
2 Ibid., vol II, p. 851.
3 Ibid., vol II, p. 212.
4 Ibid., vol II, p. 229.
5 Hermann Rauschning, *La Révolution du nihilisme* (Paris: Gallimard, 1939).
6 Cf. Paul Virilio, 'Unknown Quantity', exhibition and catalogue, Fondation Cartier pour l'art contemporain, Paris.

Chapter 2 The Invention of Accidents

1 Let's not forget that it was intensive use of powerful computers that allowed the human genome to be decoded, facilitating, by the same token, the fatal emergence of the genetic accident.
2 Organized by the European Organization for Nuclear Research (CERN), the London Institute and the Gulbenkian Foundation, the conference, 'Signatures of the Invisible', was held in Lisbon in Autumn 2002, with the participation of John Berger and Maurice Jacob.
3 Here's an example of scientific recklessness: on 13 July 2002 came the announcement of the creation in a laboratory of a synthetic poliomyelitis virus, a disease almost totally eradicated today. Robert Lamb, president of the

American Society of Virology, expressed his fears of shortly seeing terrorists perfect biological weapons of this kind.

Chapter 3 The Accident Argument

1 The International Civil Aviation Organization (ICAO) approved the creation of a worldwide protocol of aviation insurance covering risks of war which would fill the gap in the case of partial or total withdrawal by private insurers, following the attacks of 11 September. Cf. *Le Monde*, 18 June 2002.
2 Francesco di Castri, *L'Ecologie en temps réel* (Paris: Editions Diderot).
3 'La CIA au service de Hollywood', in *Le Journal du Dimanche*, 16 June 2002.
4 Arthur Miller, speaking about his book, *Echoes Down the Corridor: Collected Essays*, 1944–2000 (London: Penguin, 2001).
5 'L'Insécurité, programme préféré de la télé', in *Libération*, 28 April 2002.
6 Karl Kraus, *Cette grande époque* (Paris: Rivages, 1990).
7 Ibid.
8 Cf. H. G. Wells, *The War in the Air* (London: Penguin 2005), first published 1908.

Chapter 4 The Accident Museum

1 *Le Monde*, 24 February 2001.
2 Chronique de Jacques Julliard, in *Le Nouvel Observateur*, 30 January 2002.
3 Aristotle, *Physics* (Oxford: Oxford University Press, 1996), Section IV, Part C, TIME.
4 Gaston Rageot, *L'Homme standard* (Paris: Plon, 1928); a work more or less contemporary with Paul Morand's *L'Homme pressé* (Paris: Gallimard, 1941).

5 Pierre Barthélémy, 'Les astéroïdes constituent le principal risque natural pour la terre', in *Le Monde*, 28 June 2002.
6 Patrick Mauriès, *Cabinets de curiosités* (Paris: Gallimard, 2002).

Chapter 5 The Future of the Accident

1 Friedrich Nietzsche, *The Birth of Tragedy*, tr. Douglas Smith (Oxford: Oxford World's Classics, 2000) p. 100.
2 Ibid.
3 Henri Atlan, *La Science est-elle inhumaine?* (Paris: Bayard, 2002).
4 Victor Hugo, *Choses vues* (Paris: Gallimard, 2002). Madame Swetchine was a Christian Democrat and friend of Brother Henri Lacordaire.
5 Karl Kraus, op. cit.
6 Werner Heisenberg, *Physics and Philosophy: The Revolution in Modern Science* with Edward T. Heise (HarperCollins, 1962).
7 Karl Kraus, op. cit.
8 Henri Atlan, op. cit.
9 'Hermann docteur ès fraude', in *Libération*, 26 October 1999.
10 'Le savant fou', in *La Croix*, 8 August 2002.
11 Fabienne Goux-Baudiment, Edith Heurgon, Josée Landrieu (eds), *Expertise, débat public* (Paris: L'Aube, 2001).
12 Following the atomic strategy known as 'du faible au fort' (from the weak to the strong) which, with France's nuclear strike capacity, justified enlargement of the concept of deterrence between national states, we saw the launch, in 1990, of the strategy 'du faible au fou' (from the weak to the mad), to confront the problems of nuclear proliferation. Cf. Ben Cramer, *Le Nucléaire dans tous ses Etats* (Paris: Alias, 2002).

Chapter 6 The Expectation Horizon

1 'Programme sans couleurs pour les Verts,' in *Libération*, 8 May 2002.
2 Karl Kraus, op. cit.
3 Cf. supra, no. 1, p. 37.
4 Paul Valéry, op. cit., vol. II.
5 Ibid.
6 Brigitte Friang, *Regarde-toi qui meurs* (Paris: Le Félin, 1997).
7 Karl Kraus, op. cit.

Chapter 7 Unknown Quantity

1 Victor Hugo, op. cit.
2 Sigmund Freud, *Civilization and its Discontents* (Penguin Books, 2004), pp. 5–6. *Des Unbehagen in der Kultur*, first published in 1930.
3 Sylvie Vauclair, *La Chanson du soleil* (Paris: Albin Michel, 2002).
4 'Une loi pour sauver la nuit noir', in *Le Figaro*, 3 June 2002.
5 Robert Antelme, *L'Espèce humaine* (Paris: Gallimard, 1979).
6 Maurice Blanchot, *L'Entretien infini* (Paris: Gallimard, 1969).
7 Ibid.
8 Ibid.
9 Ibid.
10 'Time is the Accident of Accidents', Aristotle, op. cit, IV.
11 Victor Hugo, op. cit.

Chapter 8 Public Emotion

1 Thierry Wideman, 'La dissuasion nucléaire', in *Les Cahiers de Mars*, autumn 2003.
2 Philippe de Felice, *Foules en délire, extases collectives* (Paris: Albin Michel, 1947).

3 In September 2004, l'Espace Paul Ricard in Paris presented an exhibition-screening of Sandy Amerio's work entitled 'Storytelling Communities of Emotion'.
4 Philippe Pons, 'Le plaidoyer de Sadako Ogata pour la sécurité humaine', in *Le Monde*, 6 January 2004.
5 'Forvarsnihilism', a movement run by the Socialist Youth of Sweden.

Chapter 9 The Original Accident

1 Martin Rees, *Our Final Century* (London: Arrow Books, 2003), p. 148.
2 Ibid., p. 121.
3 Once more, we cite Aristotle: 'Time is the accident of accidents,' *Physics*, IV.
4 Franz Kafka, 'The City Coat of Arms', in *The Complete Stories*, New York: Schocken Books, 1995.
5 Jacques Isnard, 'Le sénat autorise Monsieur George W. Bush en matière d'armes nucléaires antibunker,' in *Le Monde*, 18 September 2003.
6 Hervé Morin, 'Une coulée de métal pour plonger au centre de la Terre', in *Le Monde*, 17 May 2003.
7 Denis Delbecq, *Libération*, 18 May 2003.
8 Martin Rees, op. cit. p. 32. A trace of Solly Zuckerman, an expert of long standing to the British government, appears hauntingly in W. G. Sebald's book, *On the Natural History of Destruction*, tr. Anthea Bell (London: Hamish Hamilton, 2003) p. 19.
9 Svetlana Alexievitch, *La Guerre n'a pas un visage de femme*, French tr. G. Ackerman and P. Lequesne (Paris: Presses de la Renaissance, 2004) p. 393.
10 Jacqueline Giraud, 'L'arme bactériologique: un boomerang', in *L'Express*, 25 March 1969.
11 Ibid.
12 Ibid.

13 Cf. Louis Morice in the broadcast, 'Thalassa: Escale en Crimée' ('Les damnés de l'atome', in *Le Nouvel Observateur*, 14 May 2004).
14 Jean-Pierre Vernant interviewed by Roger-Pol Droit, 'Les Jeux antiques étaient entièrement intégrés à la religion', in *Le Monde*, 21 August 2004.
15 Alain Loret, 'Inéluctable dopage?' in *Le Monde*, 24 August 2004.
16 William Bourdon, president of the association Sherpa, 'Forcément coupables', in *Le Monde diplomatique*, September 2004.
17 Locked up in Alderson Prison in the USA for having broken into a nuclear missile base in Colorado in 2002, in order to express her opposition to the plans to invade Iraq, Sister Carol Gilbert, a nun belonging to the Dominican Order, sent this message for the third anniversary of 11 September 2001: 'Let each of us express our rejection of death and act in favour of life,' *La Croix*, 8 September 2004.

Chapter 10 The Dromosphere

1 Ursula Le Guin, 'Direction of the Road' in *Buffalo Gals and Other Animal Presences* (London: Victor Gollancz, 1990), pp. 84–91.
2 Benoît Hoquin, 'L'arbre qui tue', in *Le Monde*, 13 April 2004.
3 Victor Hugo, op. cit.
4 Paul Virilio, *Negative Horizon*, tr. Michael Degener (London and New York: Continuum Books, 2005).
5 Martin Rees, op. cit, p. 8.
6 Ibid., p. 26.
7 'La recherche selon Bush et Kerry', in *Le Monde*, 17 September 2004.
8 Interview with Jonathan Randal in *Sud-Ouest Dimanche*, 12 September 2004.

9 In a letter to Jean Paulhan dating from 1943, Antonin Artaud wrote: 'The more time advances, the further away we get from the measure of time and its notion, as also from that of space, the more our consciences approach the infinite and the eternal, in a word, that intuitive and contemplative life where all the great mystics and all the saints have communicated with God.' *Œuvres* (Paris: Gallimard, 2004).

10 Henri Laurens, in an interview with José Garçon and J.-P. Perrin, in *Libération*, 2004.

11 Thomas Ferenczi, 'Analyse 2004', in *Le Monde*.

12 Paul Virilio, *L'Espace critique* (Paris: Christian Bourgois, 1984).

13 *Sud-Ouest*, autumn 2004.

14 Philippe Pajot, 'Terres géantes d'autres soleils', in *Le Monde*, 3 September 2004.

15 Sébastien Fath, '*Dieu bénisse l'Amérique!' La religion de la Maison Blanche* (Paris: Le Seuil, 2004). See also 'Billy Graham Speech September 14, 2001' online: http://www.usamemorial.org/sept11063.htm.

16 For instance, thanks to traders in quest of swift surplus values, capital whizzes around the planet several times a day. Warren Buffet (who has the second biggest fortune in the world) compares financial tools that allow such profits to 'weapons of mass destruction'. Cf. *Le Journal du Dimanche*, 3 October 2004.

Bibliography

Alexievitch, Svetlana, *La Supplication*, tr. G. Ackerman and P. Lorrain, Lattès, 1998.

Alexievitch, Svetlana, *Voices from Chernobyl*, tr. Keith Gessen: Picador, 2005 *(La Guerre n'a pas un visage de femme*, tr. G. Ackerman and P. Lequesne, Presses de la Renaissance, 2004).

Antelme, Robert, *Human Race*, tr. Jeffrey Haight and Annie Mahler, Northwestern University Press, 1998 (*L'Espèce humaine*, Gallimard, 1979).

Arendt, Hannah, *The Origins of Totalitarianism*, Harcourt, 1976.

Aristotle, *Physics*, Oxford University Press, 1996.

Atlan, Henri, *La Science est-elle inhumaine?*, Bayard, 2002.

Attali, Jacques, *Économie de l'Apocalypse*, Fayard, 1995.

Blanchot, Maurice, *L'Entretien infini*, Gallimard, 1969.

Cramer, Ben, *Le Nucléaire dans tous ses Etats*, Alias, 2002.

Dupuy, Jean-Pierre, *Pour un catastrophe éclairé*, Le Seuil, 2002.

Ellul, Jacques, *The Technological Bluff*, Eerdmans, 1990 (*Le Bluff technologique*, Hachette, 1988).

Felice, Philippe de, *Foules en délire: extases collectives*, Albin Michel, 1947.

Freud, Sigmund, *Civilization and its Discontents*, Penguin, 2004 (first published 1930).

Friang, Brigitte, *Regarde-toi qui meurs*, Le Félin, 1997.

Friedrich, Jörg, *The Fire: The Bombing of Germany, 1940–1945*, tr. Allison Brown, Columbia University Press, 2006.

Gimpel, Jean, *The End of the Future: The Waning of the High-Tech World*, tr. Helen McPhail, Greenwood, 1995, (*La Fin de l'avenir*, Le Seuil, 1992).

Hatchuel, Armand in Fabienne Goux-Baudiment, Edith Heurgon, Josée Landrieu (eds), *Expertise, débat public*, L'Aube, 2001.

Heisenberg, Werner, *Physics and Philosophy: The Revolution in Modern Science*, with Edward T. Heise, HarperCollins, 1962.

Hugo, Victor, *Choses vues*, Gallimard, 2002.

Jonas, Hans, *The Imperative of Responsibility: In Search of an Ethics for the Technological Age*, University of Chicago Press, 1985.

Kraus, Karl, *Cette grande époque*, Rivages, 1990.

Lagadec, Patrick, *La Civilisation du risque*, Le Seuil, 1981.

Mauriès, Patrick, *Cabinets of Curiosities*, Thames and Hudson, 2002 (*Cabinets de curiosités*, Gallimard, 2002).

Monod, Théodore, *Sortie de secours*, Seghers, 1991.

Morel, Christian, *Les Décisions absurdes*, Gallimard, 2002.

Nietzsche, Friedrich, *The Birth of Tragedy*, tr. Douglas Smith, Oxford World's Classics, 2000.

Picon, Antoine (ed.), *La Ville et la Guerre*, Éditions de l'Imprimeur, 1996.

Poliakov, Léon, *La Causalité diabolique 1. Essai sur l'origine des persécutions*, Calmann-Lévy, 1980.

Rauschning, Hermann, *The Revolution of Nihilism: Warning to the West*, Kessinger, 2005 (*La Révolution du nihilisme*, Gallimard, 1939).

Rees, Martin, *Our Final Century*, Arrow Books, 2003.

Sebald, W. G., *On the Natural History of Destruction*, tr. Anthea Bell, Hamish Hamilton, 2003.

Stiglitz, Joseph E., *Globalization and Its Discontents*, W. W. Norton & Company, 2002.

Valéry, Paul, *The Collected Works of Paul Valéry, Volume 10: The Outlook for Intelligence*, tr. Jackson Mathews, Princeton University Press, 1989 (*Cahiers* 1894–1914, Gallimard, 1987).

Vauclair, Sylvie, *La Chanson du soleil*, Albin Michel, 2002.

Vié Le Sage, Renaud, *La Terre en otage: gérer les risques naturels majeurs?*, Le Seuil, 1989.

Virilio, Paul, *Cybermonde: La Politique du pire*, interview with Philippe Petit, Textuel, 1996.

Virilio, Paul, *A Landscape of Events*, tr. Julie Rose, MIT Press, 2000 (*Un paysage d'événements,* Galilée, 1996).

Virilio, Paul, *Unknown Quantity*, tr. Chris Turner, Thames and Hudson, 2003 (*Ce qui arrive,* Galilée, 2002).

Virilio, Paul, 'Unknown Quantity', exhibition catalogue, tr. Chris Turner, Fondation Cartier pour l'art contemporain, Paris, March 2003.

Wells, H. G., *The War in the Air*, Penguin, 2005, (first published 1908).

Wiener, Norbert, *God and Golem, Inc.: A Comment on Certain Points Where Cybernetics Impinges on Religion*, MIT Press, 1966.

Index

www.ingramcontent.com/pod-product-compliance
Lightning Source LLC
Chambersburg PA
CBHW072109250125
20788CB00003B/13

* 9 7 8 0 7 4 5 6 3 6 1 3 9 *